P9-DTY-549

Lean TPM

A Blueprint for Change

Lean TPM

A Blueprint for Change

Dennis McCarthy and Dr Nick Rich

ELSEVIER
BUTTERWORTH
HEINEMANN

AMSTERDAM • BOSTON • HEIDELBERG • LONDON • NEW YORK • OXFORD
PARIS • SAN DIEGO • SAN FRANCISCO • SINGAPORE • SYDNEY • TOKYO

Elsevier Butterworth-Heinemann
Linacre House, Jordan Hill, Oxford OX2 8DP
200 Wheeler Road, Burlington, MA 01803

First published 2004

British Library Cataloguing in Publication Data
A catalogue record for this book is available from the British Library

Library of Congress Cataloguing in Publication Data
A catalogue record for this book is available from the
Library of Congress

ISBN 0 7506 5857 6

For information on all Elsevier Butterworth-Heinemann publications
visit our website at http://books.elsevier.com

Printed and bound in Great Britain

Contents

Foreword _____

If European manufacturers are going to succeed and prosper for the future, they need to drive the necessary productivity and hence competitiveness by unlocking the potential of their biggest asset: their people.

I am delighted, therefore, to support my colleagues, Dennis McCarthy and Nick Rich, in the publication of this book. I have been a long and passionate advocate of TPM since my visits to Japan in the early 1990s. As an example, I fondly remember being told by the Japanese Managing Director of a recognised World Class manufacturer and TPM practitioner that:

> ... in the '50s and '60s we had 'M' for Manufacturing. In the '70s we had 'IM' for Integrated Manufacturing. In the '80s we had 'CIM' for Computer Integrated Manufacturing ... He paused for a moment and then added For the remainder of this decade and 2000 and beyond, my company is going to be pursuing 'CHIM': Computer Human Integrated Manufacturing ... We have decided to re-introduce the human being into our workplace!

Today, some 12 years later, my interpretation of that powerful message is that it certainly represents a challenge for all of us to develop and harness people's skills – not just hand/operational skills, but team-working and problem-solving skills – to deliver the technological advantage that increasing speed, high precision and automation of equipment and processes will apparently give us. Perhaps naively, I think of 'Lean' as increasing the velocity from customer order to the receipt of the goods and payment of invoice, through the elimination of waste in all its forms. Fortunately, most of the human race abhors waste. So, if the agenda is explained clearly regarding why we need to continuously improve to drive out waste and we involve people throughout the organisation from the bottom up in that goal, then the company or enterprise stands a better chance of survival and indeed sustainable growth. Lean does not work without highly reliable and predictable machines and processes. You only get this result through highly reliable and predictable people who take ownership and responsibility for their actions at the workplace.

TPM is a powerful tool in the right environment and in the right hands because, as a shop steward once told me, TPM stands for Today People Matter. The tools and techniques are the easy bit. The challenge for the modern day manager is to create an environment where people want to continuously challenge and change 'the way we do things around here'.

That's why this book will help you with these challenges – so here's to a Totally Productive Read!

Peter Willmott
Chairman
SUIKO-WCS

Preface

The purpose of this book is to set out how the convergence of Lean Thinking and Total Productive Maintenance presents a comprehensive blueprint for business-led change. This also sets out how leadership and strategic thinking are an important part of the recipe for successful and sustained improvement. This is not intended as a comprehensive guide to Lean Thinking or TPM techniques, but a user manual on how to deliver business benefits from their application. It is assumed that the reader already has some awareness of the basics of these world class manufacturing techniques. The book is organised to be read from cover to cover. For those of you with specific needs, below is a summary of the chapter content. For those planning a new improvement process, Chapters 1 to 5 will be the most important. For those wanting to take their existing activities to a higher level, Chapters 2, 3, 5 and 6 will help to identify where you are on the journey and assess what comes next.

Chapter 1 sets out the common reasons for improvement initiative failure and how these can be addressed through Lean TPM. Chapter 2 provides summary details of the origins of Lean Thinking and TPM and sets out the route map used by companies who have successfully achieved world class levels of excellence. Chapter 3 sets out the foundation for identification of hidden losses/wastes linked to accountabilities by management level. Chapter 4 sets out the leadership roles of the change team to involve all functions within a single change agenda. Chapter 5 describes the challenges to be overcome to achieve the foundation stabilisation level, targeting internal processes at customer value and establishing the capability to achieve zero breakdowns. Chapter 6 describes the leadership challenges to break through the glass ceiling and achieve market-leading capability. It covers strategic issues, organisational flexibility and optimisation processes. Chapter 7 considers issues of sustainability at management and operational levels. Finally, Chapter 8 provides a summary of the key changes incorporated in the Lean TPM change blueprint.

Acknowledgements _____

For me, the starting point for this book goes a long way back to my early career in food wholesaling/distribution. The company I worked for serviced corner shops at a time when supermarkets were taking their market away from them. As a junior member of the management team we restructured and refined the business model so that we fared better than most. I wouldn't have missed the experience but I have never forgotten how it feels to be in the wrong place at the wrong time. Too many improvement programmes stop short of addressing flaws in the basic business model. Shopfloor improvement can only go so far.

Another important lesson learned, as I managed projects in Logistics, Supply Chain and then Manufacturing was the benefits projects gained through the removal of barriers between functions. High levels of cross-functional understanding have an immense impact on increasing the pace of improvement. The text book organisational models with their clinical role descriptions do little to explain the dynamics of how each function works let alone how they could work better together.

When I arrived at the start of my TPM journey, I found the answer to successful improvement in the form of the concept of hidden losses. These losses explain the reality of day-to-day problems which are accepted as inevitable. These appear to be independent occurrences rather than recognisable and therefore predictable patterns of failure. Not only does TPM identify that there is a pattern, in fact there are six of them, it also provides the tools to reduce/eliminate them. In almost every case, the solution involves closer cross-functional co-operation. With this comes understanding, innovation and the potential for accelerated business improvement. There is something new to learn from every improvement programme so it is difficult to list the organisations and individuals that have contributed to the ideas which are incorporated in this book. I apologise in advance for those who are not named individually. I acknowledge a giant debt to both Peter Willmott and David Buffin who have shared their knowledge with me over the years. Hopefully they will recognise the principles and values that they hold dear as well as some of the practical solutions to change and improvement that we developed together. Simon Billett stands out as a consultant who has shaped my thinking particularly in the Quality Maintenance areas but I have been lucky to work with a great team at WCS whose support over the years has helped me enormously.

No small credit must also go the JIPM who collated their experiences to produce a TPM process which has become an essential part of the world class recipe. My involvement with two study tours to Japan to visit their exemplar sites has been a corner stone of my education in how improvement can become continuous. In pulling the material together I have very much appreciated the opportunity to work with such an experienced practitioner of Lean as Nick Rich. This has significantly smoothed the process of blending together two sets of experience to produce a robust and practical improvement tool. Last but not least, thanks go to Karen, my wife, who has had to suffer intense bouts of writing 'the book' for almost 2 years.

Dennis McCarthy

First, I must thank my family for their indulgence and kindness. Fiona, my wife, has suffered the most and is now a great children's entertainer for Daniel (the demanding 6 year old) and Joshua (whose ability to climb anything in the house usually gets him in trouble). Both I fear are destined to work in engineering in some way. I'd like to thank Dennis for all his efforts and the countless cups of coffee in various exotic locations alongside Britain's motorway network in putting this project 'to bed'.

At Cardiff Business School, I must acknowledge the team at the Lean Enterprise Research Centre for all their support and for making working life so entertaining! There is seldom a dull moment at the college. Professors Roger Mansfield (Director), Peter Hines (LERC) and Dan Jones deserve special mention for allowing me the freedom and 'air cover' to pursue my studies. I must acknowledge my sponsors who have lived through many of the 'turning points' described in this book. These sponsors range from helicopter manufacturing to contact lenses, from lipsticks to nuclear fuel rod assemblies and from crafting crystal glass to chemical reagent manufacturing. I am lucky to have such a group of people to work with and for. They know who they are, they don't need a mention and will have to settle for a freebie copy of the book as a token of my gratitude!

I owe a lot to John Moulton (Network Events) who is always an inspiration and has a phenomenal ability to define good avenues for research – he would have made a great academic. The trade union

Amicus must be acknowledged especially the efforts of Sir Ken Jackson and Lynn Williams. Amicus has been instrumental in generating true management–union partnerships and promoting TPM for productivity and quality improvement in this country. Sir Ian Gibson is another modern 'industrial hero' to whom this country and the engineering professions owe a big debt. Rarely attracting praise, although deserving, are the members of HM Dept of Trade and Industry (including their regional support networks and various 'industry forum'). Talents such as those of Francis Evans and Nigel Goulty have played a tremendous role in promoting the 'manufacturing cause' and long may they do so.

Finally, I could not finish an acknowledgement without mentioning Toyota and supply chain that have taught me and given me so much. I will be eternally grateful. This group of businesses continues to grow, diversify and reinvent itself. Companies like Denso and Aisin (the pioneers of what we now know as TPM) deserve special mention for 'blowing my mind' with their relentless pursuit of optimisation. Finally, and not least, the JIPM sensei have for decades guided TPM implementation programmes and, despite their demanding manner, have provided the pillars and the keys to push businesses to the ultimate levels of operational efficiency and competitive effectiveness. To everyone I thank you very much.

Nick Rich

Abbreviations

ABC	Classification approach (criticality)
ABCD Goals	Accidents, Breakdowns, Contamination and Defect Analysis
CANDO	Cleaning, arranging, neatening, discipline and order in workplace organisation (aka 5S)
CBM	Condition Based Monitoring
CFM	Cross Functional Management
D2D	Door to Door OEE measurement
EM	Early Management
F2F	Floor to floor OEE measurement
FLM	Front Line Management
JIPM	Japanese Institute of Plant Maintenance
JIT	Just In Time
NVA	Non Value Adding activity
OEE	Overall Equipment Effectiveness
OTIF	On Time In Full customer deliveries
PPM	Parts per Million defects
QC	Quality Control
QDC also QCD	Quality, delivery and cost performance objectives
QFD	Quality Function Deployment design approach
R&D	Research and Development
RCM	Reliability Centred Maintenance
S2C	Supply Chain OEE measurement
SMED	Single Minute Exchange of Dies (Quick Changeover of machinery)
SPL	Single Point Lesson Instruction document
TCO	Total Cost of Ownership
TPM	Total Productive Maintenance/Total Productive Manufacturing
TPM5	Fifth Bi-annual European TPM Forum
TPS	Toyota Production System
TQM	Total Quality Management
VOC	Voice of the Customer performance expectations
VSM	Value Stream Mapping analysis
WCM	World Class Manufacturing

1

The business of survival and growth

Let's face facts – most manufacturing businesses are under pressure to compete and to extract a profit from what they convert. At the heart of the competition is the need to survive, to grow, and to capture the benchmark position for their industry. So it is frightening to think that many manufacturers don't have a business strategy and most struggle with how to manage change effectively (Brown, 1996). These processes are further complicated because markets and competitors don't stand still (Hill, 1985). They are constantly changing and adding more uncertainty and confusion for managers. Add these problems together and you have a recipe for crisis management 'Western style' and all the necessary ingredients for business failure even if you have a great product, a good brand name and capable employees. Competition is an inevitable part of manufacturing today and the ability of a firm to compete is the final arbiter of the longevity of any business.

1.1 The new competitive conditions

The modern competitive conditions have generated a new 'set of rules' for manufacturers. These new rules include the provision of the highest level of customer service, the delivery of quality products in shorter and shorter lead times and product proliferation to offer variety to customers (Brown, 1996). If you take a few minutes to consider what life was like 10 years ago and compare it to now your business has probably moved on substantially. In the past you probably recorded product quality in terms of percentage defects produced during manufacturing to the modern measure of 'Parts Per Million' (PPM) levels, you probably offer more products than 10 years ago and you would probably have halved its lead times. Taking a few more minutes, you may like to contemplate the future and guess what? These performance indicators are likely to get tougher and tougher. The new rules of competition demand the effective management of the rate of change within the business and the elimination of all unnecessary waste or costs in order to provide the ultimate levels of customer service throughout the firm.

External pressures

The race to compete and to survive is a difficult one. Markets are full of pressures that increase uncertainty for managers. In today's markets managers must be able to sense and make sense of these changes if they are to make the right and timely improvement activities within the factory. These pressures come from a variety of sources including the government (laws and taxation), customers who expect ever-improving levels of service, consumer groups who inevitably seek to lower prices, competitors, parent corporations and shareholders who demand increased returns for their investments.

The new competitive conditions are far removed from those of the past and challenge strategies such that we can no longer assume that:

- Past business success is a guarantee of future survival;
- Product patents will protect a manufacturer from competition;
- Buying the latest technology will provide a means of defence against competition.

Technology or products by themselves are not enough to guarantee survival. The countries of the developing world are eager to take their place in the world economy. They know that they have the opportunity to leap-frog the traditional costly batch and queue approach in favour of more efficient low inventory, high flow and high quality operations. With the support of organisations looking for low cost supplies they have also developed the management skills and expertise to run their operations at high levels of effectiveness.

Many offshore competitors also have the advantage of governments prepared to offer advantageous tax allowances to attract inward investment. These advantages should not be overstated. They have their problems including poor infrastructure, low home demand, poor material supply and in some cases corruption. So developing countries don't have it all their own way but when these constraints are lifted, they will be even more competitive. Naturally as this happens their costs will also rise. The size of their domestic markets and therefore global consumption will increase as a result. These trends shaping the shifting sands of the future market, can be predicted without the aid of a crystal ball. The only uncertainty is when. All the pressures and opportunities, from outside

Table 1.1 Pressures for change

1 New and emerging manufacturing economies with low labour costs are attracted to mature Western markets where they can exploit their 'cost advantage'.
2 The power of the internet in purchasing materials and components on a global scale and therefore access to alternative suppliers has increased exponentially. As such, power has shifted to the customer/consumer.
3 Deregulation of world markets has resulted from international trading agreements and this has liberated trade and increased competition for manufacturers.
4 Corporations have the ability to switch production.
5 Pressure groups and lobbyists seeking to lower prices or convince the manufacturer to improve their performance in areas such as environmental management.
6 Shareholders who expect a 'year on year' improvement in the returns on their money invested and constantly compare these returns with what their money could earn elsewhere.
7 Customers expect product variety, continuously improving quality levels, lead time reduction and want their stocks reduced.

and from within, mount up to a major challenge (Slack, 1991). However, the challenge is surmountable if management can:

- Harness the intellectual capability of the complete workforce;
- Target this creativity on making better products, more cheaply;
- Achieve 'world class' manufacturing standards that set your organisation apart from the competition.

These are the goals of Lean TPM and the subject of this book.

1.2 Silver bullets, initiative fatigue and fashionable management

Throughout the 1980s and 1990s many management books and academic publications heralded new business models that would, if applied correctly, radically transform the firm into a 'world beater' capable of meeting the demands of the market and fending off competitors (Suzaki, 1987). Also during the 1980s, Japanese texts explaining certain manufacturing techniques of high performing firms were translated into many languages creating an interest in applying these techniques in non-Japanese workplaces. Many of these models were, however, to prove seductively rational but without a methodology for implementation. The techniques provided a methodology but offered no real advice concerning how to integrate them into business-wide improvement activities or what support activities would be necessary to ensure the techniques would 'stick'. Managers in many countries readily adopted both. These managers were motivated by a number of reasons: some were keen to implement and be seen to be at the 'cutting edge' whilst others grasped at these new practices as if they were lifelines and implemented change with an air of desperation to make any form of improvement. Whilst some improvements were implemented and sustained many were not and ended in failure. Such failures did little to increase the credibility of managers with employees and even customers.

1.3 Why programmes fail

Some improvement initiatives failed because they were applied in a piecemeal way – grafted in place but rejected by those people who did not select or necessarily understand 'why' change was necessary but were tasked with implementing it and working in a new way. Others failed because they were little more than 'technical quick fixes' and 'sticking plaster' solutions and many such changes were quickly reversed, failed or left as the manager jumped quickly into the next 'fix'. Examples of these programmes included attempts to compress the set up time of machines in a belief that the company would be capable of producing high levels of variety and missing the fact that the existing machinery was not capable of meeting the quality tolerances needed of it. Indeed, whilst many project managers claimed massive 'time savings' from such a programme very little tended to be added to the bottom-line profit of the firm. These 'time savings' simply vanished because batch sizes were not reduced, quality was not improved and inventory was not withdrawn from the production facility. The failure to exploit these 'point improvement activities' is unsurprising and many companies have failed to develop and operate a standardised/stabilised *production system* that forms the basis of managing effectively. Without an effective production system concepts such as 'Zero Breakdowns' are seen as unattainable, and the mechanical cost-reduction programmes are used indiscriminately and without 'against learning from experience'. As such, any change is installed piecemeal and without a supporting environment within which the practice could be sustained (Storey, 1994).

In this manner, the impact of isolated projects deteriorate after the project closed as management attention passes away to the next 'fire-fighting crusade'. These have created many disappointments, from the management, the business and employee viewpoints. What 'glittered' and promised so much at the beginning of the change programme often delivered very little, damaged the credibility of managers as business leaders and tarnished all other change initiatives within the business. As these tools worked in some companies and not others, the techniques of World Class Manufacturing (WCM) are not at fault. To isolate the real culprit we need to dig deeper.

Management outlook

In most cases manufacturing businesses are dominated by historically derived patterns of behaviour and not behaviour that meets the needs of the modern market place. These patterns, based on a past wisdom, are rarely questioned as long as they don't fail completely. As mentioned earlier, past success is no indication of future performance and when faced with a declining financial position, poor management teamwork leads to uncertainty over the need for a new approach. Individually, managers can only engage in crisis management. As financial conditions worsen, attempts to 'weather the storm' characterise a 'denial' phase. Ultimately, a new commercial model is sought only when results get so bad that they cannot be ignored. The new model frequently involves redundancy (downsizing) and a new management structure. The success or failure of the new organisation will depend heavily on collective will and reshaping the business behind a single change agenda (Figure 1.1).

Strategically, downsizing reduces the chances of long-term business survival. At the bottom of the trough, often there is no choice to do otherwise. Companies who grow also call on the collective will of their organisation to co-operate behind a single change agenda. The difference is that they have learned to use opportunity as the trigger for change rather than rely on fate. The lost opportunity is impossible to measure but it is significant nevertheless.

Ignoring the potential of growth and adopting a strategy, which matches the business to the state of current markets, is no easy option for management. It is painful, there are dozens of examples as to how this approach has resulted in 'death by a thousand cuts' as the course of the organisation is dictated by the chaos of market forces. The end game of this strategy is totally predictable. The organisation eventually falls below critical mass where the form of the business is unable to support the overheads necessary to operate in its chosen market. Merger/Acquisition

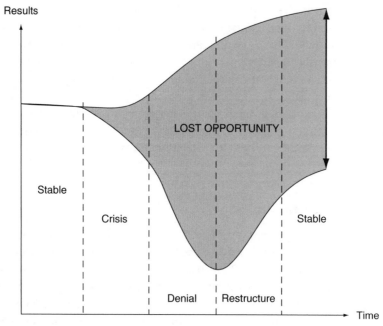

Figure 1.1 Crisis management stages: Patterns of lost opportunity

becomes the only alternative to receivership. As over 50 per cent of these fail to produce real benefits the cycle repeats itself.

Consider the following joke levelled at management by the workforce. An operations manager, on taking up a new position, found three sealed envelopes in the top drawer of his desk. The first envelope contains a page of paper that holds the following advice 'Blame your predecessors and if things don't improve open envelope 2.' This works for a while but eventually the manager opens envelope 2 seeking advice. Envelope 2 contains a similar piece of paper that urges the management to 'Restructure the organisation and if this fails then open envelope 3.' Dutifully the manager follows the advice but eventually he finds he has to open the third envelope. This letter states 'Find another job but before leaving photocopy these three letters and leave them in the top drawer of this desk for your successor.'

As with all jokes, it has a ring of truth about it and presents a view of management, which is unable to manage, and can only react to events. This depressing reality occurs all too frequently due to:

- A lack of strategic clarity and analyses of market trends;
- Inappropriate Key Performance Measures;

- Fragile Technology;
- Departmental/Silo Thinking.

Levels of participation

In the environment of top-down driven initiatives, the potential of the vast majority of problem-solvers in the firm is ignored. Here management expect 90 per cent of employees to receive rather than initiate improvements. They will be involved in the implementation rather than participate in the evolution of future working practices. This is despite the fact that these employees are the experts in their individual areas of the transformation process and are also often customers. Not only does this miss an opportunity to accelerate the pace of improvement but also it increases the risk of failure.

A recent survey of unsuccessful change projects revealed the following reasons for failure:

- The key 'influencer' left or moved positions;
- Goals too distant or too vague to engage all levels of personnel;
- Benefits/results disputed or not properly recorded;
- Insufficient training;
- Competing crisis distracts attention.

Scratching the surface of these responses using '5 why's' analysis suggests a common root cause for all these problems (Table 1.2):

Table 1.2 WHY the programme failed

The Problem: The programme failed when the champion left

1 Why? because those left behind where not motivated enough to press on with the changes needed;
2 Why? because the reasons for doing so where not compelling enough;
3 Why? because dealing with the barriers was more painful than living with the inefficiency;
4 Why? because there was not a collective will to change;
5 Why? Because not enough people shared the belief change was really necessary.

These are failures to understand how to change habits/patterns of behaviour/culture.

Cultural drivers

Culture, or 'the way we do things around here', is driven by instinctive behaviour. At the core of this is the way the brain makes decisions. The brain instinctively avoids what it perceives as discomfort/pain and seeks what it perceives as pleasurable. Many managers reading this book will begin to feel uncomfortable at this stage of the book but please persevere, as this issue is critical to the design of successful initiatives that stick and sustain improvements.

1 Your 'gut instinct', when deciding what to do next, will typically be based on patterns of past experience. It will favour things that gave you pleasure/satisfaction in the past. So employees, including managers, will have a natural tendency to avoid difficult problems – they are painful.
2 When making choices, the mind gives equal weight to ideas based on beliefs as to those based on fact. If there is a belief that 'They will never let us do that' it will not normally be subject to any test of logic. This is one of the reasons why seemingly intelligent people sometimes do stupid things.

These are emotional, instinctive responses without which we could not function. These are what helps us make the judgements that allow us to drive safely along the motorway at speed or to develop the co-ordination to 'bend it like Beckham'. Sometimes, however, learned patterns/habits can become limiting. Which is why in some organisations making changes can feel like walking in treacle.

Most manufacturers, speeding up the flow process of getting materials from 'door-to-door', especially with established operations, would recognise the importance of people involvement. Whilst certain practices can be implemented by key 'technical staff functions' like condition-based maintenance (Engineering) and advanced six sigma techniques (Quality Assurance), to be successful any improvement initiatives must impact on the transformation process which will require the operational staff to 'buy in' to the process. Even managers with newly established operations still need this form of 'mass employee' cultural development in order to secure competitive advantage through 'zero loss' manufacturing.

If, for most people, change means waiting for others to change things, this can result in learned helplessness. It reinforces a pattern best described as BOHCA syndrome (bend over here comes another). In this environment, people get used to finding their fulfilment outside of work and in the absence of information to the contrary expect change to have negative consequences.

So what can be done to tackle this important dimension of the continuous improvement agenda. Limiting patterns are reinforced day to day by relatively simple mechanisms or cultural anchors as set out in Figure 1.2. Only two of these issues are included in the scope of most improvement techniques (Organisational Structure/Internal Systems and Procedures).

To make change happen, the change process must deal with all six areas simultaneously. The 'burning platform' approach at the end of the denial phase in Figure 1.1 works by interrupting limiting patterns of behaviour by generating discomfort. This is the quickest and easiest way to get attention. It is a valid approach. Unfortunately, the 'burning platform' approach only works until the fire is put out. The pace of change driven solely by such stimulus will slow down when the pain is reduced. Like the patient with a toothache, which stops hurting in the dentist's waiting room, deciding whether to go through with the treatment. The likely return of the pain would make most people go through with the treatment. If pain is what made you act, once it is

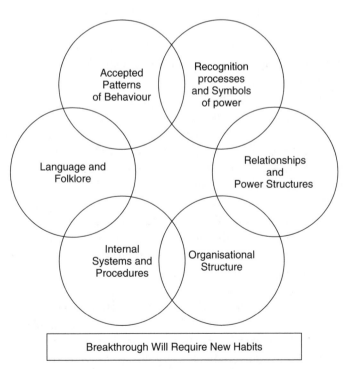

Figure 1.2 Restraints to change: Cultural anchors

gone, it is unlikely to make you adopt the change of diet/dental routine that will address the causes of decay. For change to be long-lasting, limiting patterns needs to be replaced with a more positive alternative. Longer term and sustained improvement needs to be driven by a compelling vision – a vision that the whole of the organisation can get behind. This is where successful companies score over their less successful competitors. Although the difference between success and failure is far narrower than many would imagine, this will take more than a few well-chosen motivational slogans. Most people instinctively know when they are being manipulated.

It must be backed up with a systematic process to raise standards, increase flexibility and secure outstanding performance. Only when such a vision of the future is driving the change process will the organisation have developed the collective competence to anticipate and lead the customer agenda (an outcome to which all world class organisations aspire).

1.4 The value of a compelling vision

Of all the inputs to the factory, it is perhaps this lack of people engagement that represents the greatest hurdles to improving factory performance (Brown, 1996). Without the engagement of the 'many', the problems of improving and competing will remain the prerogative and responsibility of the 'few' organisational managers within the firm. Vesting the entire future of the firm in the hands of just a few managers is quite a concern. No single manager can revolutionise the business model. Sure some managers are inspirational leaders but they also attract a following of people who deliver the changes. No single manager can hope to perform his or her planning role and also execute the huge amount of change needed to be competitive and stay ahead of the game.

By engaging workers, those people who determine factory efficiency and generate a stable material flow system, managers can be released to plan the future of the firm and to devise the key business improvement programmes that will transform business performance. An empowering vision and focus provide meaning for everyone in the organisation. In this way, they can identify with the challenge and become a part of it. If you think that this is unattainable, the next time you visit an exemplar company, observe how passionate the people are about their company. Organisations who find it difficult to achieve performance improvements seem to continuously change their business structures creating pain and uncertainty (in terms of roles and fit) for everyone.

So an appreciation of how best to engage and develop the skills of the workforce to increase the quality and quantity of improvement programmes within the door-to-door flow is highly important in the short term and also as the foundation for a sustainable improvement process.

Understanding behaviours

To express the problem in a different manner, what is needed is a means of enlisting the support of all workers in the planned change. This will only happen when each worker finds fulfilment in the process of change. The concept of individuals needing a reason for change/wanting to know what is in it for them is perfectly reasonable. As mentioned earlier with change comes discomfort. Learning new patterns of working takes effort and anyway we have an inbred preference for the status quo. So there must be something worthwhile in the change process or people will need to be dragged kicking and screaming. On the other hand, if the change is of interest, it will be a case of 'light the blue touch paper and stand well clear'. So what motivates people?

Despite the fact that research into performance-related pay demonstrates its failure to sustain increased performance, there are still those who make the mistake of trying to buy co-operation. If this stalwart of management tradition doesn't work, what does?

Research shows that in addition to our basic survival needs we are also driven to satisfy other needs such as identity, excitement, learning, and to feel valued. These needs are met in many different ways but it is what drives football fans to follow their teams across the world, backpackers to endure the discomfort/personal risk and pilgrims to search for enlightenment. They have all found a 'compelling vision' that gives them focus, meaning and a positive physiology (Figure 1.3).

Team development

Although less exotic, these 'human needs' are also met through the rapport generated by positive relationships and team development. In the workplace, this can also be provided by the development of effective team-working. In practical terms this means building the improvement process around balanced teams of five to seven people with a clear goal and the capability to achieve it.

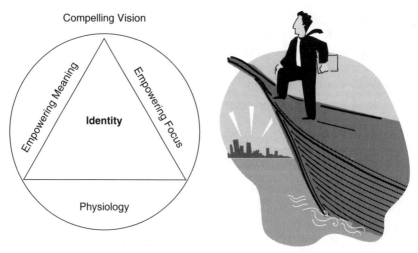

Figure 1.3 Developing a vision to engage individuals

Unfortunately, although teamwork has been a part of world class manufacturing for over 20 years, scant attention is paid to the structured development process. The development of foundation level team working does not mean sending everyone on 'outward bound' courses, although this may be useful in some circumstances.

The following three-step process sets out a proven route for establishing basic level team working:

- Step 1: Provide information to help employees to position themselves in the big picture. For example, raise understanding of the entire supply chain of the firm and how quickening the material-to-cash cycle will improve the performance of the firm;
- Step 2: Define boundaries/focus/roles for each team and involve them in identifying improvement proposals that will speed up flow. To achieve this level of continuous flow all barriers must be broken down in a logical sequence that starts with the availability and quality of each stage in the flow process.
- Step 3: Support the drive to achieve and sustain zero targets across the value chain from product design, through operations, final shipment and after-sales service of the product.

Refer back to Figure 1.2 and consider what impact this process could have on the cultural anchors in your organisation.

The outcome is the creation of cells/teams who know how to develop proactive relationships to support the application of technical skills. This is one of the outcomes that can be delivered as part of a properly structured 5S/CANDO workplace organisation process (discussed in more detail in Chapter 5). Once such basic team-working capability has been mastered it provides the foundation for the evolution of these teams to higher levels of empowerment and self-management (the complete High Performance Teamwork Development process is set out in Chapter 3).

1.5   Leading the improvement process

The value of an organisation that practices a vision-led set of management behaviours is clear. In such an environment teams can (with support) develop their ability to become self-managed. Leaders at all levels are able to delegate routine tasks to those who are nearest to the day-to-day operation and thereby create the necessary space and time for managers to direct and co-ordinate improvement efforts.

Identifying the problems facing manufacturing firms is a relatively easy process. More difficult is the process of designing the improvement programme through which employees can engage their learning and secure incremental mastery of key business processes (and by default the techniques needed). This can be likened to dialing a telephone number; the digits have to be selected in the right sequence to get the right connection. The paragraphs below set out an overview of the leadership dimensions of a successful improvement programme.

Review/Formalise current practices

The first stage in the commercial–cultural improvement model is to formalise current practices to enhance value by establishing good practices across the organisation. This removes wasted effort and provides a foundation for sharing ideas. As the initial layer of waste and inefficiency is removed, processes can also be simplified, there is also less management fire fighting and meaningful pockets of time can be released. This means that activities can be redefined and routine activities delegated. Learning therefore takes place that releases valuable specialist resources to address the next layer of waste/value. When these are brought under control, processes become simplified and in turn support further delegation.

Build rapport across functions and organisations

This process of 'horizontal empowerment' blurs the traditional boundaries between functions and levels. Properly managed, it builds rapport and better working relationships which in turn improves communication. It also reduces the well-documented constraints to growth caused by the lack of availability of skilled labour.

Talent management and personnel development are as important to future success as securing funding. Those organisations that have experience of Lean production and have engaged in autonomous maintenance will know well these issues and the benefits of correctly aligning the worker, learning and improvement activities.

The potential of these rapport-building skills extends beyond the boundaries of the organisation to include supplier relationships. Traditionally, they have been treated as 'enemies', accounting for a high percentage of manufacturing costs. This ignores the potential of supplier relationships built on releasing innovation in preference for one built on 'annual price negotiations'. This is despite the evidence that no matter how tight the legal contract, they are no substitute for a proactive working relationship.

So in many respects, the external world beyond the factory gate and modern competitive environment is more complicated than it could be because many firms have not found a methodology and business model with which to integrate and focus resources within the firm.

Learn from experience

The new 'manufacturing challenge' is therefore for managers to engage all levels of employees in building robust, dependable and flexible manufacturing processes that create a purpose-designed manufacturing system of 'delay free' material flow within and beyond the factory gate. The methodologies behind the technical, silver bullet tools and techniques are well covered and can achieve 'point benefits' for elements of the door-to-door flow but enlisting the 'hearts and minds' of an empowered workforce is the key to long-term and sustainable organisational learning that delivers results.

Encourage questions

Our desire to learn, with careful design, can be tapped into and used to support the pace of change of today's markets for manufactured products. At the heart of this learning capability lies understanding the customer and differentiating between what adds value and helps material to flow and what adds costs and waste. Teaching employees to understand and reflect upon how best to change the organisation is important to nurturing the 'learning culture'. It is no surprise that managers adopting this approach do not get upset when subordinates question their change programmes and decisions. Questions are one of the most important countermeasures to interrupt patterns of behaviour and counter 'group think'. They are to be encouraged. Those of you who have taken part in behavioural safety will recognise that questions are an important indicator of interest, learning and engagement.

Anticipate the growth stages

A recent study of the progress made by organisations on their journey to excellence shows that the learning mechanism is a key lever to developing collective capability as a means of accessing new and more productive patterns of working. These steps of this 'learning to improve' journey are summarised in Table 1.3 (p. 16) which is presented as a map of guiding principles. Once again there are no universal solutions and each solution, adopted by a company, must be designed for that company.

Progress through each milestone depends on the organisation's ability to adopt a single change agenda or improvement theme to align cross-functional improvement efforts. As each step provides a viable working environment without this common theme, it may not be possible to break through to the next level. Taking short-term benefits at each stage can present a management trap which will constrain progress and increase the risk of regressing back to the lower levels where there is less protection from the changing winds of the competitive market place.

Organisations, which progress past milestone 3, have achieved the capability to further transform their business. They will understand how the commercial value of addressing new 'zero losses' becomes clear as the skills to achieve current targets are mastered.

In this way, the last two steps are iterative, representing a practical model of the world of never-ending improvement. These 'zero loss' targets include accidents, breakdowns, specific quality defects, set up times, energy loss and a host of others we will explore later in this book.

Establish a single change agenda

A sustainable and continuously improving manufacturing system therefore can only be built incrementally. Each stage is mastered and a competence gained before moving to the next. This means mastering a basic level of competence in Process Stabilisation (see Chapter 5)

Table 1.3 The improvement journey stages and management issues

Step	Improvement tactics	Target
1	Set standards for equipment, processes and behaviours. Formalise current practices against them. Provide information and education to address recurring problems. Establish basic flow processes, remove excess inventory and control sources of dirt, dust and accelerated deterioration.	Define and maintain basic conditions. Stabilise lead times. Establish basic team-working practices. Engage all employees with the company competitive agenda.
2	Simplify and refine core activities and remove sources of accelerated deterioration to deliver the 'zero breakdown' goal. Refine flow layout and inventory/planning parameters to reflect true demand profile.	Achieve 'zero breakdowns'. Compress internal lead times. Establish high performance teamwork.
3	Transfer/delegate all routine activities to all operations teams. Refocus specialist resources to optimise value adding processes and address the causes of quality variations. Extend time between intervention (reel to reel running).	Achieve 'zero unplanned intervention'. Install low cost automation. Establish flawless introduction of new products and processes. Compress total supply lead times, remove cross-functional barriers.
4	Condition way of working to sustain process optimisation activities. Create value from reduced variation in quality and delivery and market-leading innovation.	Achieve 'no touch production'. Reduce new product time to market. Increased customer loyalty and growth.

before mastering Process Optimisation (see Chapter 6). It is important to understand the sequence and logic of production system development so that the learning/improvement process is linked to the delivery of increasing levels of manufacturing advantage. In turn, this becomes part of the mechanism to secure a compelling vision for the organisation (Figure 1.4).

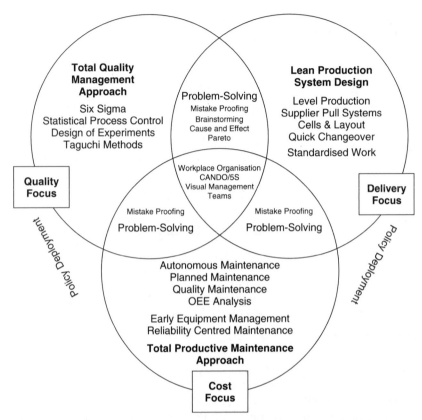

Figure 1.4 World class manufacturing techniques (Rich, 2002)

Although 'zero accidents' and 'zero breakdowns' are the prime targets during the initial stages of stabilisation, actions to secure these will also increase operations effectiveness and efficiency. Assuming that there are no major safety hot spots, when selecting the area to start priority should be given to opportunities to improve the quality of the entire linked chain of 'door-to-door' activities (both office and production area) that result in the quality of product offered to the customer (Table 1.4).

From here, the focus moves on to shortening the time from receipt of order to customer delivery and then to find new ways of offering high variety performance (in shorter and smaller lot sizes). This logic is quite easy to understand and each stage removes a level of waste and cost from the business. Obviously some investments will need to be made along the way but overall the important task of providing customer service should result in the total cost of manufacturing falling and productivity (including learning about how best to be productive) rising.

Table 1.4 The five sources of manufacturing-led competitive advantage (adapted from Slack 1991)[1]

Competitive (market) advantage	Role of business and operations management
Quality	To make things right based upon the needs of the customer.
Speed	To make things fast and in shorter amounts of time.
Delivery	To make things on time with dependable delivery times.
Flexibility	To offer product variety and have the ability to change/update the product catalogue (what is made).
Cost	To make things at the lowest cost, to enhance margins within the guideline that the product is 'fit for purpose' and meets customer needs.

So, despite the complexity of the business problems facing manufacturing management, the logic of World Class Manufacturing (WCM) performance is not that difficult to understand or build (Schonberger, 1986). In fact, clarity and consistency of purpose are essential levers in the process of securing lasting change. In this way all employees from senior managers, trade union officials, to the newest of operators and office workers can understand it. This form of incremental mastery allows each new chapter in the development of the company improvement process to be evolved with the workforce before each change theme is launched. In this manner, the management prerogative to lead change is reinforced and the application of techniques have a logic and a purpose rather than just asking employees to change without an understanding of the bigger picture. As most managers know from bitter experience this latter form of change management breeds frustration and often ends in disappointment.

1.6 Lean TPM

So far in this chapter we have set out the challenges facing manufacturing and suggested that these can be met if management can:

- Harness the intellectual capability of the complete workforce;
- Target this creativity on making better products more cheaply;
- Achieve world class manufacturing standards that set the organisation apart from the competition.

These are the goals of Lean TPM which are delivered by a process which combines the development of leadership and management best practice to secure long-lasting change. The key features and benefits of the approach is set out in the paragraphs below.

An approach based on proven business models

The combination of Lean Thinking (Womack and Jones, 1996) and Total Productive Manufacturing (Lean TPM) applies the proven business models of 'world class' manufacturing firms who have learned how to dictate the rate of change and competitiveness of their chosen markets. This incorporates a change process which is far removed from 'blank silver bullets' and whilst each company must develop their own unique approach, the model provides many proven design principles for managers and the workforce.

Challenges limiting behaviours

Lean TPM creates a potent blend of case studies, principles and techniques to challenge old patterns of behaviour and replace them with a more versatile, flexible and proactive outlook.

Techniques to develop a compelling vision

The Lean TPM approach also presents a 'future state' business model within which empowerment and learning combine to allow 'mastery' of those key customer winning criteria that mark out high performers from the 'also-rans'. Both approaches promote the hypothesis that the future of a manufacturing business depends upon its employees and the need for learning and innovation in current practices.

Promotes 100 per cent participation

The Lean TPM approach promotes improvement at the point of activity. It secures the engagement of the entire workforce in change and innovation in terms of thinking about how work is conducted, identifying waste and workers as the source of new ideas, new ways of working and sustainable improvements. Whilst some businesses display posters stating that *'Employees are our Number One asset'*, Lean TPM companies create a compelling vision to truly engage their employees and reinforce the involvement of the workforce by getting them involved rather than clinging to mantras.

Supports organisational learning

Combining efficient design with a focus for organisational learning provides access to and new more effective ways of working especially given the power of TPM. TPM has the proven power to break through the learning barriers that have prevented a meaningful optimisation of the manufacturing process and up-skilling of operator teams to engage in greater diagnostic improvements related to the assets they control.

Incorporates a comprehensive loss measurement system

One further aspect of the Lean TPM approach to which we have alluded during this chapter but not yet explored is a very powerful Lean TPM measurement system. This measurement system goes beyond the traditional measures of manufacturing. It provides visibility of previously hidden management losses in such areas as planning processes, new product development and technology losses, and unmet customer needs. The measurement system tells each manager how far the business has progressed and whether improvement activities are generating an increased competitive capability for the firm. To date, there are many examples of 'kamikaze improvements' that have glittered and released absolutely no benefits to the firm or its customers. The Lean TPM approach is not so forgiving. It is not a blunt measure nor is it one that can easily be 'manipulated' as so many previous measures of 'world class' performance have been.

At the basic level there is the analysis and trend information that relates to a single asset or cell. This has been referred to as the Overall Equipment Effectiveness (OEE) measure (Nakajima, 1986) or the 'floor to floor' level of analysis (see Chapter 3 for calculation). This aspect of the Lean TPM measure shows how well a machine/cell is managed. The disadvantages of OEE measures include the ignoring of the chain of machines that supply (or take products from) the asset and form the internal production chain. Hence, the 'door-to-door' measure include these linkages. Finally, the highest level of control and the level at which 'manufacturing' can be exploited, as a means of competitive advantage

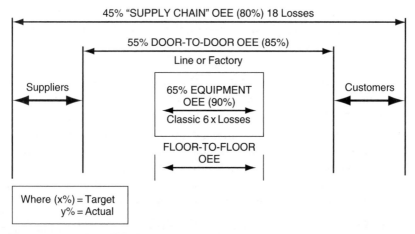

Figure 1.5 LEAN TPM measures

is the 'supply chain OEE'. These are interrelated measures, giving different levels of management analysis and trend information, covering the entire production system. These measures can be understood by all managers and can be used to target improvement/optimisation efforts and rid the firm of waste. We will return to the use of these measures and their role in the optimisation process in later chapters (Figure 1.5).

The measurement system therefore allows for proper navigation of the firm starting with the optimisation of individual assets, the optimisation of the chain of assets in the factory that form the production sequence and finally the overall performance of the firm and its selected supply chain design.

1.7   Summary: The foundation for a better improvement model

The modern competitive world calls for the management of key customer processes. These processes include:

- Basic activities that are combined to just simply meet the demands of the market and can be thought of as 'conditions for trade';
- Key processes that 'delight' customers and are 'market beating'.

For most operations this includes at least the management of quality and delivery processes and the mastery of these key processes cannot be given to an individual with the firm whatever the title of the manager. As such the quality assurance manager cannot, by himself or herself, guarantee the quality of the factory. Instead, the manager is dependent upon the actions and decisions of the employees engaged in purchasing, operations, maintenance and also logistics to name but a few. The same is true of the delivery process and also that of cost management and cost reduction. No single manager has the ability to truly influence the production system. However, as a group of key stakeholders, who each shares a part of the value creating stages from door-to-door, it is possible by concerted effort to make substantial and radical improvements. Interestingly enough, from a management perspective, the modern competitive environment calls for the mastery of the door-to-door processes of quality, delivery and cost management (Figure 1.6). These processes are what the customer values and differentiate those organisations from whom the customer will or will not place an order or custom. So 'world class' performance and highly effective manufacturing businesses recognise this 'lateral' nature of customer wants even though the business is not structured in that manner the business is managed as a system and through high levels of cross-functional management. Such an approach also demands that each manager must learn about the business and the needs of other managers in the campaign to raise customer service.

So, in the modern competitive market place we have argued that the 'new art' of management, given what we have discussed so far, concerns how to compete effectively in an ever uncertain and competitive market through the management of cross-functional resources

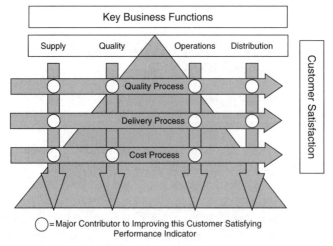

Figure 1.6 Cross-functional management

including people, knowledge and processes. We have also identified the importance of mastering key processes and taking a door-to-door perspective (from goods inwards to the shipping bay and beyond). This demands the company must:

- Design *efficient* products, services, operations and systems;
- Progressively raise the *effectiveness* of all elements of the supply chain;
- Guide the development of the complete company team towards the delivery of a clear and compelling business strategy.

The reason why this works in exemplar organisations is that it covers all three of these in parallel.

What we have presented in this chapter is a 'blueprint for change'. The mandate for change already exists for every manufacturer. The Lean TPM blueprint, whilst specific to each firm that uses it, has an unforgiving power that is commercial and embodies the process optimisation and 'zero losses' needed to compete effectively and to engage change in a meaningful manner that is focused and understood by every employee.

Bibliography

Brown, S. (1996) *Strategic Manufacturing for Competitive Advantage*. London: Prentice Hall.

Hill, T. (1985) *Manufacturing Strategy*. Basingstoke: MacMillan.

Nakajima, S. (1986) *Introduction to TPM*. Portland OR: Productivity Press.

Rich, N. (2002) Turning Japanese? PhD Thesis, Cardiff University.

Schonberger, R. (1986) *World Class Manufacturing: The Lessons of Simplicity Applied*. New York: Free Press.

Slack, N. (1991) *Manufacturing Advantage*. London: Mercury Press.

Storey, J. (1994) *New Wave Manufacturing Practices*. London: PCP Press.

Suzaki, K. (1987) *The New Manufacturing Challenge*. New York: Free Press.

Womack, J. and Jones, D. (1996) *Lean Thinking*. New York: Simon and Schuster.

2
Lean TPM

2.1 Achieving the right balance

At the TPM5 conference (November 1997) the biannual conference of European TPM practitioners, Professor Daniel T Jones addressed the conference delegates on the topic of Lean Thinking and TPM. His observations were that although Just-In-Time is an accepted concept, most industries still scheduled work through departments in batches, worked to forecast and sold from stock, had long lead times, high buffers and poor quality detection. These are key target areas for 'Lean production' and the lean enterprise business model. To the casual observer the lean approach has a different emphasis to the classic TPM focus on equipment reliability. There is some overlap, but together these cover twelve different target areas. So why would a recognised leader of 'Lean Thinking' be talking at a TPM conference?

The common thread is that both TPM and Lean Manufacturing highlight areas of historically accepted or hidden wastes (Womack and Jones, 1996). Despite their different origins, progress with either depends upon sensitising the organisation to recognise wasteful behaviours and practices; in effect, to create such a heightened sensitivity to these 'wastes' that each employee detects these issues as abnormal and takes appropriate actions to eliminate them. Such an approach makes employees quite intolerant to other organisations that still maintain old business models and have not yet engaged in this form of waste elimination. In the case of TPM the root cause of this waste is a short-term perspective that tolerates poor reliability. The root cause of Lean wastes is optimising parts of, rather than, the total value stream. TPM companies have always channelled improved effectiveness to increase customer value, but Lean Thinking helps to sharpen the definition of value. Lean Thinking has always sought reliable processes, but TPM provides the route map to zero breakdowns and continuous improvement in equipment optimisation.

The penultimate slide in Dan Jones's presentation showed the potential gains from Lean as reducing:

- Throughput time and defects by 90 per cent;
- Inventories by 75 per cent;
- Space and unit costs by 50 per cent.

Overall, this potential to double output and productivity with the same head count at very little capital cost could equally be presented as the potential of TPM. Both Lean and TPM have evolved in parallel from their early concepts and are converging towards a common goal. But who cares? As long as there are benefits all ideas are welcome. To understand what these are, it is worth taking a brief journey through the origins of both Lean Thinking and TPM.

2.2   The origins of Lean Thinking

Sakichi Toyoda and his son Kiichiro, the family that founded the Toyota Motor Corporation, began to produce weaving looms and then cars in the 1930s. The approach taken by the family was to engage in a variant of flow production that later matured to become known as the *Toyota Production System* (TPS) and has more recently become known as 'lean production' (Monden, 1993). At the heart of the manufacturing system was an attention to using simple machinery that automatically stopped, and assembly lines that could be stopped by operators, when a defect was detected (a system known as *Jidoka*). In this manner, no defective products would be passed forward to internal customer operations (Shingo, 1981).

In the West, large batch sizes and the responsibility for quality inspection being the role of a specialist department meant that defects could move and be hidden in buffers only to generate interruptions downstream as defects were filtered. Other factors conspired against the development of the mass production system at Toyota, not least the lack of natural resources and large amounts of capital to fund investments in large-scale and dedicated technology. To counteract the lack of resources Toyota engaged a production system that did not rely upon forecasts for each department but used a pull system (Ohno, 1988b). Under the pull system, parts actually needed by internal customer operations or customers are made (called *Just-in-Time*). As such production strictly controlled and standardised inventory buffers deliberately disconnect operations. In this way the movement of production materials from a supplier to a customer operation created a replenishment order. The basic pull system was later supplemented with a deliberate approach to level the workload of each production area (called *Heijunka*). It was not until after the Second World War that Taiichi Ohno (Toyota's chief engineer) compiled these practices to form the lean Toyota Production System that exists today (Womack and Jones, 1996).

Ohno-san was a man with a vision and the architect of the Toyota Production System (TPS). His intent was to introduce a production system of high-variety production in small volumes. Such an approach was therefore completely at odds with the Western passion for large batch sizes, dedicated and expensive technology and forecasting all operations. He commenced the production system for engine manufacturing (where he had served his time as a Toyota employee) before extending it to vehicle assembly and later to include all Toyota suppliers (during the 1970s). In effect, Toyota now had a total 'pull system' network of materials supply that allowed instant availability of materials and a system that worked to replenish (pull) what had been consumed rather than pushing huge batches through operations to meet estimated forecasts.

Figure 2.1 shows how a pull system works with a limited amount of finished products held to allow immediate customer satisfaction. Upon consumption the inventory level drops and this causes an internal order (kanban card) to be returned to the drilling operation. The drilling

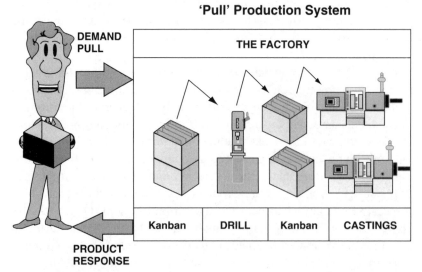

Figure 2.1 Pull system

line then takes products, from a controlled kanban stock, and drills the hole in the product. As the inventory between the drilling line and the diecasting operations falls this places an order on the diecasters and the return of a kanban card. The diecasters replenish the orders and so on through the manufacturing process.

The traditional approach is to use a push system and starts with a forecast or order provided by the customer for products at a certain date in time (Rich, 1999). Let's say the 27th of the month. If it takes 2 days to drill the holes in the product then that means the products must arrive on the 25th of this month and if it takes 3 days to diecast then the products must be launched on the 22nd of the month. So the physical manufacturing process starts with the release of materials in the diecast section on the 22nd of the month and the pushing of the products to the drilling station to meet the deadline of the 25th and then the push through drilling to the customer. That differentiates a push from a pull system at the most basic level. Obviously any slippage in the push system, such as a machine breakdown, will stop the flow and immediate satisfaction of the customer whereas under the pull system, the standard kanban inventory provides customer service and affords some protection from internal process disruptions.

To support the pull production approach, Ohno and Toyota had also engaged in widespread quality management processes using employee involvement and had focused these activities on eliminating 'waste' from factory activities (Ohno, 1988b). The great quality guru, Dr Edwards Deming, echoed these principles in his works on industrial quality management (Deming, 1986). In this manner, Toyota had, rather than concentrating on making machines process quicker (and engage in batch production with high buffers), changed the emphasis towards the redesign of the production system. Such a design allowed the elimination of all the poor features of mass production that generated excessive production costs and slowed the flow of materials in the factory. These forms of factory waste were identified as seven key aspects of production management – 'the seven wastes'. By concentrating upon the elimination of waste, the amount of value-adding time improves as materials do not sit in buffers or contain defects. As a result flow performance improves and the cycle between paying for materials, conversion and sale of the materials is compressed to improve the operational and financial performance of the firm (Table 2.1).

After an extensive study funded by the Western automotive industry (IMVP Study), the performance advantage of Japanese Automotive Manufacturers was recorded in the publication, *The Machine That Changed The World* (Womack et al., 1990). It was this publication that

Table 2.1 The Toyota seven wastes

The waste of OVER-PRODUCTION where vast amounts of products are made in batches and simply 'dumped' into finished goods or work-in-process and result when there is a mismatch between customer demand for products and the ability of the production system to make to that demand. This is one of the greatest problems with mass production and the reliance upon large batch sizes.

The waste of UNNECESSARY INVENTORY where the results of over-production and other 'unimproved' constraints means that inventory is simply held awaiting an order in the belief that future orders will come.

INAPPROPRIATE PROCESSING is another waste that results from a mismatch between the processes needed to make a product and the processes that are in place. In this manner, many firms use very sophisticated machinery to manufacture simple products that would be best produced using 'simpler' and less expensive technology. Typically, in the West, large sophisticated machines with high processing speeds tend to be 'pumped full' of production in order to ensure a 'pay back' for the asset and keeping such machines occupied with work inflates batch sizes and generates inventory (two forms of waste).

UNNECESSARY TRANSPORTATION is a further form of waste concerning the movement of materials around a factory from the receiving back to the shipping bay. This activity can consume many hours and involve many kilometres of transport (with each activity offering the potential for product damage).

UNNECESSARY DELAY concerns the simple 'dwelling' time as products are ready to be converted but sit waiting. For much of factory time, materials will be 'idly hanging around' in an uncontrolled manner.

UNNECESSARY DEFECTS is the production of materials (that consumes value-adding time) but have to be reworked or scrapped. In this way valuable capacity is lost forever – you cannot reclaim it even by working overtime. So imagine the problem of large batch sizes, long travel distances and, hidden within these batches, defective products!

UNNECESSARY MOTION occurs when the production process is poorly designed and operators engage in stressful activities to handle materials. This is an unusual waste (ergonomics), but as claims for repetitive strain injury rise many firms are facing large settlement fees from employee claims and solicitor bills.

first used the term 'lean production' to describe a new form of manufacturing developed by Toyota and adopted by most Japanese assemblers. The estimated advantage, resulting from the factory benchmarking process, was concluded to be a Japanese advantage of 2:1 in productivity terms and nearer 100:1 in quality of vehicle build (Figure 2.2). The basis of this 'manufacturing mastery' was found to be a tightly integrated and synchronised manufacturing and supply system that, due to the lack of stock buffers in the total system of material flow, was termed 'lean production' (Womack et al., 1990). To put it another way, the Japanese producers could make products in half the time of the West and enjoyed the benefits of near-perfect materials entering the vehicle build process (measured in terms of PPM defects rather than percentages).

In a later study of automotive component manufacturing plants in the UK and Japan (Andersen Consulting, 1993), five factories, of eighteen in the survey, were deemed to be 'world class' and managed to generate high levels of quality and productivity simultaneously (Figure 2.3). These factories were Japanese but not all Japanese managed to achieve the 'world class' status. Interestingly, when a line of 'best fit' is drawn between the non-world class companies (Japanese and British) it shows an old engineering adage – that you can have productivity but you will generate poor quality or that you can have slower production but good quality. This ability of 'world class'

Indicator	Japanese in Japan	All Europe
Performance		
Productivity (hours/car)	16.8	36.2
Quality (Defects/100 cars)	60	97
Layout		
Space (sq.ft/car/year)	5.7	7.8
Inventory (sample 8 parts)	0.2	2.0
Size of Repair Area		
(% Assembly Hall)	4.1	14.4
Workforce		
% in Teams	69.3%	0.6%
Suggestions/Employee	61.6	0.4
Absenteeism	5%	12.1%
Training of New Production		
Workers (hrs)	380.3	173.3

Figure 2.2 The benchmarking findings (Womack et al., 1990)

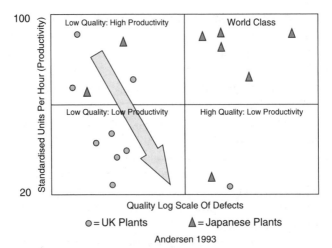

Figure 2.3 Andersen benchmarking findings

companies to break this trade-off using lean production principles firmly established the lean approach to manufacturing as an alternative design to that of traditional mass production.

In 1994, the study was repeated (same products) but included a wider sample of countries involved (nine countries and eighty-one manufacturing sites). The survey results, whilst less dramatic than its predecessor, revealed an 'under-performance' by Western firms and a dominance of Japanese firms (and emulators of lean production). This dominance encompassed all measures of 'world class' performance (quality, delivery, costs and people metrics) and the producers all exhibited the same type of manufacturing system; they were all 'lean producers' and these studies firmly founded the lean model as an alternative to the 'ails and weaknesses' of traditional mass production designs (Andersen Consulting, 1994). These surveys also confirmed the competitive power of lean systems especially for manufacturers in very competitive and harsh markets and the ability of lean producers to maintain the highest levels of customer service with low levels of inventory buffering.

Lean Thinking: Beyond auto production

The power of the lean production system design did not escape the attention of companies in other industrial sectors and many Western manufacturers, in industries far removed from car production, began to adopt aspects of the lean production model and to enjoy performance improvement. These businesses included aluminium-converting companies, aerospace businesses and general manufacturers. The importation of the lean production model, using its logic rather than a simple copying of techniques, was termed 'Lean Thinking' and was popularised by Womack and Jones (1996) in a book of that title. The book examined over fifty cases of emulation and demonstrated 'before and after' comparisons of performance improvement. The authors also set out five basic pillars of lean thinking that, when implemented in order, generated the foundation for high performance lean manufacturing (Table 2.2).

From this time onwards, the terms 'lean production' and the 'lean enterprise' entered popular management terminology and have continued to redefine the model of the post-mass organisation. Lean production and the emulation of lean systems is now practised in a wide variety of industrial sectors, by large and small companies, and is rapidly being transferred to the supply bases of companies 'going lean'.

2.3   The origins of TPM

The planned approach to preventative maintenance was introduced to Japan from the US in the 1950s. Seiichi Nakajima, of the Japan Institute of Plant Maintenance (JIPM), is credited with pioneering the development of the approach through the stages of Preventative (Time-based) Maintenance, Productive (Predictive/Condition-based) Maintenance and then into Total Productive Maintenance. This work is inextricably linked to the maintenance systems designed and perfected by 'Toyota family' companies including Denso and Aisin Seiki (Nakajima, 1988). The JIPM went on to identify the following five critical success factors for delivering benefits from TPM:

- Maximise equipment effectiveness;
- Develop a system of productive maintenance for the life of the equipment;
- Involve all departments that plan, design, use or maintain equipment in implementing TPM;
- Actively involve all employees from top management to shopfloor workers;
- Promote TPM through motivation management: autonomous small group activities.

Table 2.2 The pillars of lean thinking

1 Understand Value in terms of '**WHAT**' the customer wants to buy and what provides customer satisfaction/customer service. This stage includes understanding the wastes in the current production system that stop or delay the process of information and material movements to provide ultimate levels of customer value. A general 'rule of thumb' used widely suggests that less than 5 per cent of the time materials spend in a production system is spent having value added (converted ready for sale rather than delayed, transported etc. which simply adds costs and no real value for customers).

2 Identify the Value Stream and the internal activities undertaken within the firm that converts a customer order into a fulfilled order and the activities associated with generating new products for customers. Once you understand how you manufacture and design products you can improve the process and from here you can begin to work with the wider value stream (suppliers and customers) to eliminate all the wastes between companies involved with satisfying customers.

3 Make Products Flow is the third pillar of lean thinking and involves keeping materials and information moving so that materials 'flow' to customers without delay or interruption. Stocking materials for very long periods of time reduces stock turns and this inflates costs and ties up huge amounts of capital in materials that are not being sold for a profit.

4 Pull Production at the Rate of Consumption is used when it is not possible to completely flow products to customers (due to the number of customers, short lead times, the needs of your technology and batch sizes or other constraint). Under this rule, where it is not possible to flow production, a buffer must be deliberately designed to allow customer orders to be fulfilled from a carefully managed stock point. In this way, it is possible to maintain customer service by later production and finishing processing 'pulling' out the work they need to complete orders from this buffer point. For advanced forms of lean production it is possible to have many small buffer points that are used to directly link internal customer and supplier production operations and allow customer orders (removed from finished goods stocks) to completely pull work through the factory. This is known as the kanban system at Toyota and allows instant availability of products and short lead times simultaneously.

5 Seek Perfection in every aspect of the business and its relations with customers and suppliers is the final pillar and rule of lean thinking. Here, the authors stress the use of problem-solving teams of operators, managers and inter-company teams to squeeze out the last remaining elements of waste and non-value-added activity.

This focus on total involvement and motivation management is in recognition of the need to establish a realistic perspective towards equipment at all levels. The truth uncovered through TPM is that if equipment fails to deliver its 100 per cent potential, it is due to some physical phenomena which can be identified, brought under control, reduced and possibly even eliminated. The JIPM identified *six categories of equipment loss*:

1 Breakdowns due to equipment failures;
2 Set up and unnecessary adjustments;
3 Idling and minor stops;
4 Running at reduced speed;
5 Start-up losses;
6 Rework and scrap.

They also identified that the *main reasons for such losses* are:

- Equipment condition is poor;
- Human error/lack of motivation;
- Lack of understanding of how to achieve optimum conditions.

It is a short step from this finding to recognising that 'zero breakdowns' can be achieved through establishing a good standard of equipment condition provided users/maintainers of equipment develop and apply practices which minimise human error and improve early detection of potential failure.

The technical case for zero breakdowns relies on the fact that in a production environment, where the level of minor quality defects are one source of advance warning, early detection is possible with around 85 per cent of components. Much of this is best detected through constant monitoring by those involved in the production process. For those components where it is only possible to detect failure (such as light bulbs), given the same operating conditions, the useful life of the component will be predictable. Creating optimal conditions will not only prolong the life of such components. It will also make it easier to predict the most efficient servicing cycle.

The pragmatic case for zero breakdowns rests on the fact that, in reality, most catastrophic failures are due to lack of or over-lubrication. Another large element of failures is due to poor operation/finger trouble. If you accept breakdowns as inevitable, such patterns of working are reinforced.

A common inhibitor to establishing optimum conditions is the scattering of dust and dirt, which results in accelerated deterioration. The build up of dirt on equipment can also mask the early signs of failure. This is why 'cleaning is inspection' is a frequently quoted TPM mantra. Finally, the most common root cause of equipment, which fails to work properly, is lack of understanding. In extreme cases, this can result in calls for alternative technology. If you have not learned to master the current technology, what makes you think that you will do better with a new set of problem issues. This is the reason for another important TPM mantra 'restore before improve' (Tajiri and Gotoh, 1992).

The top-down management role

It is clear that the maintenance department can't deliver high levels of equipment effectiveness alone. Only management have the opportunity to set and enforce standards/policy to address equipment losses. It is they who define priorities and allocate resources. The top-down role is to define priorities and set standards, and must provide recognition when they are achieved such that the right behaviours are reinforced. Within classic TPM there are five key areas or pillars where standards need to be established by management. These are defined by the *five principles of TPM*:

1 Adopt improvement activities designed to increase the overall equipment effectiveness by attacking the six losses;
2 Improve existing planned and predictive maintenance systems;
3 Establish a level of self-maintenance and cleaning carried out by highly trained operators;
4 Increase the skills and motivation of operators and engineers by individual and group development;
5 Apply early management techniques to design in low life cycle costs by creating reliable and safe equipment and processes, which are easy to operate and maintain.

Bottom-up role

TPM recognised the importance of operator involvement developing a multi-disciplined Operator/Maintainer team approach to increase capability and break down traditional barriers. Experience showed that this could only be achieved over time as an evolutionary process. This supports the 5S/CANDO workplace organisation process which was enhanced to provide a seven-step process leading to self-directed maintenance carried out by operators (also known as 'the seven steps of autonomous maintenance').

The first four steps of autonomous maintenance provide the mechanism for raising equipment condition to a level where zero breakdowns is possible. Supporting this are four corresponding planned maintenance steps to guide the standardisation and simplification of maintenance activities. This stepwise process has the effect of raising the capability of production, maintenance and supervision and of releasing specialist resources to focus on the next development stage. That is process optimisation, a key part of the 'proactive maintainers' role once breakdowns are brought under control.

Integrating top-down leadership and bottom-up delivery

The stepwise shop floor TPM process builds capability, increases production capacity, and improves ownership/morale. The TPM master plan is the mechanism used to co-ordinate the company-wide learning curve and focus on the delivery of business goals. Initially, this master plan approach was developed to guide the evolution of cultural change. Later this framework was refined to guide the 3–5 year business transformation process and produce operations capable of world-leading levels of performance (Akao, 1989).

The evolution of continuous improvement

The evolution of Total Productive Maintenance into company-wide TPM or Total Productive Manufacturing occurred because of the recognition that as processes became more reliable, the level of management fire fighting reduced. Managers had time to manage, creating opportunities to adopt more productive ways of working. Therefore it is important to note that Lean TPM does not mean lean staff and redundancies – quite the opposite. The Lean TPM approach is concerned with maximising the value of managers and operations staff, not removing them from the business but harnessing these skills for business growth (Womack and Jones, 1996). As the impact of more productive roles and ways of working was achieved, support functions became subject to the application of TPM. As such, pillars for Administration, Safety and Environment, Quality Maintenance, Product Development and Focused Improvement were developed to provide further reinforcement to the 'total' approach to productive manufacturing. In this way it was able to provide a means of aligning improvement across the organisation.

The earliest example of TPM used in Europe to deliver world class performance was in Belgium where Volvo won the PM prize for their work in the paint shop in Ghent. It was adopted quickly by a number of automotive companies as they took massive action to try and catch up with Japanese levels of quality and productivity. In the UK, its potential really became recognised in the early 1990s. Although thought of, incorrectly by some managers, as a way of reducing maintenance head count, where it was implemented correctly it produced benefits for individuals as well as company profitability.

A trade union leader presented another landmark statement. Writing at the time, Sir Ken Jackson, then General Secretary of the AEEU commented 'We recognised the value of TPM some years ago. We saw then that TPM could enable manufacturing and services to become the best in the world. Unlike TQM, which was conceptually sound, but patchy in outcome, TPM offers a new and invigorating approach. Involving everyone from shopfloor to boardroom, TPM is a team-based and freshly focused tool for success.' Such an endorsement by one of Britain's most progressive trade unions is important and underlines the 'growth potential' associated with the development of operations personnel and increasing their skill sets to take on higher levels of self-management in the factory.

2.4   In summary, what does Lean TPM offer?

Lean thinking tools improve the design *efficiency* of transformation processes providing the potential to deliver greater customer value with less effort. This includes frameworks to identify 'customer winning' patterns of operation and opportunities to secure competitive advantage. TPM tools improve the *effectiveness* of the transformation process (i.e. dealing with the reasons why things don't go to plan). This includes frameworks to release capacity, increase control and repeatability. The implementation process is designed to change attitudes, develop capability and increase cross-company collaboration. Below are examples of how they work together to provide a holistic approach to continuous improvement driven by progressively removing inhibitors to and tuning the complete supply chain (Table 2.3).

Using Lean TPM to challenge current thinking and clarify business drivers is an important part of the implementation route map. This provides a practical way of making business drivers and their links with continuous improvement visible. All business functions can then

Table 2.3 The benefits of Lean TPM

Measure	Impact of TPM	Impact of lean thinking
Productivity	Reduce need for intervention Reduce breakdowns	Reduce non-value-adding activities, increase added value per labour hour
Quality	Potential to reduce tolerance Control of technology Reduce start-up loss	Highlight quality defects early
Cost	Reduce material, spares	Lower inventories
Delivery	Zero breakdowns predictability	Shorter lead times, faster conversion processes
Safety	Less unplanned events Less intervention Controlled wear	Less movement, less clutter Abnormal conditions become visible easily
Morale	Better understanding of technology More time to manage	Less clutter Closer to the customer Higher appreciation of what constitutes customer value
Environment	Closer control of equipment Less unplanned events/human error	No 'over-production' Systems geared to needs not theoretical batching rules

target continuous improvement of business drivers, making it part of the day-to-day routine. The result is an increase in the range of options available to a business when responding to economic change and a key management resource supporting the drive to sustain competitive advantage.

2.5 Tackling the hidden waste treasure map

The most important leverage to support successful waste reduction is clarity of accountabilities. Accountabilities drive priorities, so if accountabilities don't change progress will be short-lived. The concepts of equipment Floor-to-Floor (F2F), Door-to-Door (D2D), and Supply

Chain (S2C), link together to form a complete value stream measurement system. This performance management framework links top-down priorities with bottom-up shift team delivery. Accountabilities are set relative to the decision horizon, which each level of the organisation controls (short, medium and long term). Success at each level is, therefore, linked to the effectiveness of activities in the other two (Figure 2.4).

The design of this approach is based on the measurement of effectiveness metric using the following definitions:

1 *Availability Losses* relate to issues which prevent the task from starting;
2 *Performance Losses* relate to issues which reduce the output of the conversion process once under way;
3 *Quality Losses* relate to issues that reduce the quality of the output.

Each of these losses contains:

• *Unplanned* losses (breakdown, minor stops, rework);
• *Systematic or Planned* losses (set-up, reduced speed, start-up losses).

Applying these definitions to each of the three levels of effectiveness creates a treasure map containing eighteen hidden loss categories or wastes (see below). Like a map, this provides a guide to the territory

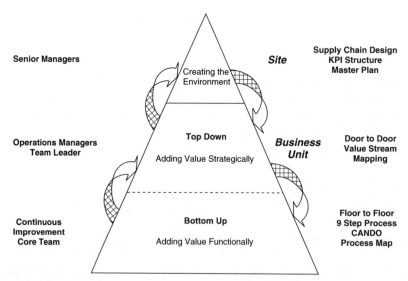

Figure 2.4 The organisational structure of Lean TPM

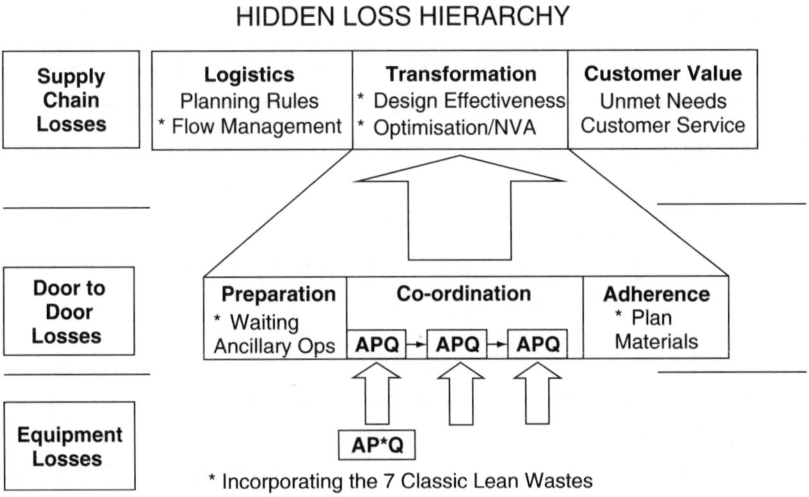

HIDDEN LOSS HIERARCHY

Supply Chain Losses	Logistics Planning Rules * Flow Management	Transformation * Design Effectiveness * Optimisation/NVA	Customer Value Unmet Needs Customer Service

Door to Door Losses	Preparation * Waiting Ancillary Ops	Co-ordination APQ → APQ → APQ	Adherence * Plan Materials

Equipment Losses	AP*Q

* Incorporating the 7 Classic Lean Wastes

Figure 2.5 Measuring Lean TPM losses and wastes

rather than a detailed replica of the actual landscape. The result is a framework, which combines the seven Lean wastes with the six TPM losses (Figure 2.5). It also identifies other hidden loss areas relating to Logistics, Co-ordination, Creating Customer Value and Design Processes.

With clear accountabilities, when the team succeed their value to the organisation can be recognised and savings established. This reinforces the right values, beliefs, profit improvement focus and the central concern of all manufacturers – to raise customer service. These losses are discussed in more detail in Chapter 3.

2.6 The integrated route map

Case studies of successful companies show that they pass through a recognisable series of stages on their journey to world class. The paragraphs below set out the journey through this implementation master plan. It is also based upon the work of the Lean Enterprise Research Centre at Cardiff Business School (one of the world's recognised experts in this field). The route map illustrates the changing leadership and organisational learning challenges presented as each layer of waste is 'peeled back'. This helps management to recognise and prepare for those challenges, minimising the risk of inertia and loss of direction.

Continuous improvement strategy development

Pilot/Launch: Continuous Improvement Strategy Development (Understanding Value the Voice of the Customer)

Aim: To achieve senior management consensus concerning the future vision, assess the current benchmark/potential and provide the hands on model and experience necessary to develop realistic and achievable route map to world class levels of performance.

- Management awareness of competitive conditions and competitor analysis to focus the appropriate response and direction of change needed by the firm.
- Refine measurement process, carry out scoping study, gap analysis and accountabilities.
- Macro mapping to identify base case and improvement potential.
- Confirm strategic vision/goals and define in detail the operation needed to deliver those goals.
- Align short, medium and long-term goals under a single change agenda/master plan.
- Establish change infrastructure including recognition systems and motivational levers.
- Develop 12 month plan in detail including logistics of releasing people for training and to support the programme.
- Apply Lean TPM to pilot value streams/centres of excellence.
- Company-wide engagement with 5S/CANDO (see Chapter 5).

Milestone 1: Roll out cascade (Integrating the Internal Value Stream)

Aim: To establish the company-wide 'best practice recipe' for low inventory, high flow, and stable operation.

Bottom-up: First Line Management (FLM) led

TPM tasks:

- Communicate commitment and gain 'buy in'
- Establish teams and build plans
- Awareness raising and TPM education
- Effectiveness measurement
- Equipment condition and criticality analysis
- Understand and reduce scattering of dust and dirt (contamination)

- Asset care/maintenance process
- Formalise operations start-up, steady state, close-down routines

Lean tasks:

- Understanding the value stream and system of production
- Process flow value mapping, identification and elimination of waste
- Flow alignment/cell creation
- Stabilise inventory levels
- Non-Value-Added (NVA) removal

Top-down: Management Pillar Champion Led

Formalize best practice standards to guide progress towards master plan goals such as zero breakdowns or class 'A' planning system that ensures all bills of materials, routings, inventory and forecasts are correct to achieve 100 per cent customer service.

Exit criteria (How to identify when the capability to progress further has been reached)

- All employees involved with high levels of FLM ownership.
- Performance gaps are assessed to focus improvements. Improvement activities undertaken to deliver stable operations.
- Factory contains showcases of good practice.
- Technical maintenance, industrial engineering and quality management records are reviewed, streamlined and compiled to form an effective management reporting system.
- The costs of poor performance are assessed and tracked to demonstrate the relationship between the ownership of improvement and waste (cost) reductions.

Milestone 2: Refine Best Practice (Make Product Flow)

Aim: To 'lock in' the recipe for low inventory, high flow operation delivering zero breakdowns and self-managed teamwork.

Bottom-up: First Line Management (FLM) led

TPM focus – F2F losses

- Simplify and consolidate maintenance tasks to reduce technical judgement. Introduce single point lessons and review process documentation for efficiency. *Note*: A single point lesson (SPL) is a single piece of A4 documentation that contains all the information needed to learn or conduct an operations/maintenance task safely and efficiently. These documents often include flowcharts to represent procedures, machine drawings, and/or digital photographs to show the procedure simply.

- Achieve real empowerment/shift beliefs by engaging in further rounds of structured problem-solving including detailed analyses of cause and effect relationships together with the inclusion of mistake proofing devices to prevent problem recurrence.
- Achieve cross-functional shared ownership of assets between operating shifts. Also the establishment of common practices such that innovation is shared, through standard process documentation, as the platform for new improvements.
- Clarity concerning operations 'trouble map' so that improvements become focused on removing greater levels of waste.

Lean focus – D2D losses

- Compress internal value stream by co-location of machinery to form cells where possible.
- Focus on quality of product to identify and react to the production of defective materials. Defect detection prompts problem-solving activities by teams. Focus on quality is intended to raise productivity and lower costs whilst also reducing batch sizes and inventory needed to support the operation.
- Common measures of performance displayed in factory areas including measures of morale, safety, quality, delivery and cost reductions achieved by the teams.
- Integration of support departments especially those engaged in tooling manufacture and departments affecting the new product introduction process.
- Also new policies affecting the allocation of maintenance spares, location and control of these items. Also strategies for sub-assets are determined effectively (i.e. repair versus replace cycles).

Top-down: Management Pillar Champion Led

- Improve underpinning training and increase diagnostic tools available to improvement personnel.
- Align activities and share results internally through presentations and copying of practices between areas with similar issues.
- Transfer roles, new skills sets and change job descriptions. Establish competency records to show employees who have received training

(internal and external) and what level of achievement they have reached (i.e. trained, capable of training or expert).

- Integrate shopfloor team outputs into planning process and capital projects.
- Prepare vision alignment to make use of increased capability/lead time.

Exit criteria (How to identify when the capability to progress further has been reached)

- No recurring problems and greatly extended time between maintenance interventions.
- Stable operating conditions and predictable usage of consumable and spares items.
- Routine self-managed maintenance and operations combined within a structure of SPLs that are reviewed by teams to ensure they represent good safe working practice and are efficient. An SPL is a single A4 paper document that explains all the major features of a process or task so that anyone can follow and understand it.
- Recognition structure for contribution of teams, suggestions for improvements, training and performance (safety and efficiency).
- Factory environment containing high levels of visual management (colour coding) and maximum use of communication boards containing key information about company and area performance.
- Impromptu and formalised problem-solving groups.
- Documentation of production system and integration of the logic and techniques of the production system integrated with employee induction on 'on the job' training.

Milestone 3: Build capability (Extend Flow Systems)

Aim: To identify the recipe to release the full potential of the current operation and build the foundations to match and exceed future customer expectations. This includes raising standards to deliver outstanding performance and the flexibility to deliver world-leading levels of performance.

Bottom-up: First Line Management (FLM) led

TPM focus

- Identify zero target priorities to optimise customer value.
- Define and control parameters to optimise manufacturing progress.

- Reduce labour intervention with machinery and need to adjust machinery to maintain quality performance. Significant progress towards 'no touch' production releasing operations staff to engage in project work and problem-solving with a greater technical content.
- Raise shopfloor team competence/capability to deliver self-management.

Lean focus

- Lead time reduction activities and time compression from the value chain of operations.
- Statistical process control engaged at all critical assets to detect, predict and control production. Self-recording by local area teams.
- Quick changeover between products and strategy to achieve 'single minute' or 'one touch' set-ups for maximum product variety to be manufactured at each stage of the production process.
- Production teams and support staff engage in the analysis of competitor products.
- Self-certification of suppliers and integration of strategic suppliers with operations system. Including levelling of demand for materials to avoid traditional problems associated with poor forecasting accuracy of production schedules.

Top-down: Management Pillar Champion Led

- Focus on competitors and how to 'step change' internal production system to levels needed to compete effectively in the future. Extension of thinking to cover next 5 years and not just the current day.
- Setting the competitive agenda and predicting and promoting future customer needs.
- Promoting innovation and establishment of key cross-functional business improvement initiatives including a focus on the supply chain and integration of supplier businesses needed to enhance material flow. Integration and development of engineering services and spares providers.

Exit criteria (How to identify when the capability to progress further has been reached)

- Have identified critical optimisation targets (commercial, operations and technical) and making progress towards them.
- Maintained zero breakdowns.
- Established a clear product/service 'innovation stream' strategy in place capable of achieving customer leading performance.

- Focus shifted from internal improvement to include external partners.
- Technical focus shifts to external scanning for asset innovations rather than internal correction. Engineering staff to ensure that they are included in any future asset specifications at the capital expenditure and procurement stages. Feedback by improvement teams to this knowledge base is routine. This includes intelligence about improvements needed for the next generation of assets.

Milestone 4: Strive for Zero (Perfection)

Aim: To change the competitive landscape and set the future customer agenda for products and services.

Bottom-up: First Line Management (FLM) led

TPM focus

- Optimisation of asset performance.
- Flawless integration of new technology (including commissioning).
- Increased focus on knowledge management/cross-project learning.
- Flexible production without labour constraints.
- Zero quality defects and losses.
- No boundaries between functions.
- Use of P-M analysis Reliability-Centred Maintenance and Condition-Based Monitoring procedures to optimise operations.

Lean focus – Supply Chain Loss

- Establishment of fully integrated supply chain, speed of supply and pull system of materials with customers and suppliers. Integration of pull system with customers creates a form of dependency and affords some protection from competition.
- Re-engineered supplier evaluation processes and integration of suppliers with business strategy sharing, key change programmes and widespread development/best practice sharing. Rationalisation of suppliers to eliminate excessive numbers of alternative suppliers or to create 'systems' suppliers providing ranges of products or configuring entire product sub-systems rather than just offering components.
- Development of concurrent product development with suppliers. Sharing of resources between companies including 'resident engineers' at supplier factories.

- Common logistics systems and use of common logistics providers (use of JIT deliveries and milk rounds for product collection).
- 'Right sizing' of tooling and assets to meet life cycle needs of product. Simplification of machinery (avoidance of procuring unproven technology). Maintenance and life cycle cost of ownership routines fully developed and integrated with product costing systems.
- The weight loss of converting raw materials to finished products examined to find methods of reducing the amount of conversion necessary. Integration of suppliers with near net weight programmes.
- Full integration and standardised process for the introduction of new products in short cycle times. Mass customisation of products to allow a logical variety of products offered to the customer but with minimum disruption to the production process.
- Environmental effectiveness of the organisation and its technology prioritised as a key competitive capability.
- Business engages in extensive networking with other businesses in non-related fields to seek out new innovations.

Top-down: Management Pillar Champion Led

- Consolidation of optimisation gains and focus on steps to control competitive agenda.
- Internal improvement resources and trained personnel directed to assist suppliers and customers with improvement activities.
- Increased deployment of business costs to shopfloor teams. Increased integration of teams with the execution of business policy deployment. Increasing responsibility of factory teams to propose key changes and present these to management as annual themes or key projects to deliver market success.

Exit criteria (How to identify when the capability to progress further has been reached)

- Delivering industry leading standards of delivery of customer value.
- Established strategy to disrupt current competitive landscape and control the rate of change for other organisations in the same sectors and market segments.
- World class standards of product innovation and customisation strategies.

2.7 Summary

Both Lean and TPM have evolved in parallel from their early concepts and are converging towards a common goal (Rich, 2002). Both are company-wide approaches and not narrow sets of techniques. They have both achieved significant results by delivering practical solutions to different business issues. 'Lean Thinking' has tools to design efficient supply chains (Womack and Jones, 1996). TPM has tools to improve supply chain effectiveness (Willmott and McCarthy, 2000). The combination of these approaches improves both operational efficiency and organisational effectiveness. Without this focus it would be all too easy to make improvements but not to convert this efficiency into cash by lowering inventory buffers in a controlled manner. In this respect a staged approach to implementation must be adopted to avoid 'kamikaze improvements' that generate more heat than improvement. To achieve improvements it is imperative that the approaches are not defined narrowly and that a cross-functional management (CFM) infrastructure is created to ensure the benefits of the change programme can be exploited properly.

The implementation route map provides the change process to align accountabilities and progressively ratchet up operational capability. Such a route map helps to co-ordinate the application of Lean Thinking and TPM tools and techniques to secure continuous improvement in business performance in terms of Quality, Cost and Delivery.

Bibliography

Akao, Y. (1989) *Hoshin Kanri*. Portland, OR: Productivity Press.

Andersen Consulting (1993) *The Lean Enterprise Benchmarking Report*. London: Andersen Consulting.

Andersen Consulting (1994) *The Second Lean Enterprise World-Wide Benchmarking Report*. London: Andersen Consulting.

Deming, W. E. (1986) *Out of the Crisis*. Cambridge, MA: Center for Advanced Engineering MIT.

Monden, Y. (1983) *Toyota Production System*. Atlanta: Institute of Industrial Engineers.

Nakajima, S. (1988) *Introduction to TPM.*, Portland, OR: Productivity Press.

Ohno, T. (1988a) *JIT for Today and Tomorrow*. Portland, OR: Productivity Press.

Ohno, T. (1988b) *Toyota Production System: Beyond Large-scale Production*. Portland, OR: Productivity Press.

Rich, N. (1999). *TPM: The Lean Approach*. Liverpool: Liverpool University Press.

Rich, N. (2002) Turning Japanese? PhD Thesis, Cardiff University.

Shingo, S. (1981) *A Study of the Toyota Production System*. Tokyo: Japan Management Association.

Tajiri, M. and Gotoh, F. (1999) *TPM Implementation*. New York: McGraw Hill.

Willmott, P. and McCarthy, D. (2000) *TPM: A Route to World Class Performance*. London: Butterworth Heinemann.

Womack, J., Jones, D. and Roos, D. (1990) *The Machine That Changed The World*. New York: Rawson Associates.

Womack, J. and Jones, D. (1996) *Lean Thinking*. New York: Simon and Schuster.

3

The change mandate: A top-down/bottom-up partnership

3.1 The Lean change mandate

Chapter 1 concluded that the future of manufacturing business depends on its employees and that efficient design with a focus for organisational learning provides access to more effective ways of working. Chapter 2 deals with tools for improving effectiveness as well as efficiency. This chapter focuses on how employees at all levels can work together to make that happen. The three sets of tools set out in Figure 3.1 provide the delivery mechanism to achieve lasting change by engaging all employees (that includes managers and shopfloor workers!).

Lean TPM principles provide a framework to identify the gaps in current model/working practices and also to help identify a more effective future model. Identifying your weaknesses is relatively straightforward but revising your business model takes time, care and must be understood by all in the factory if it is to succeed (Standard and Davis, 1999). As such the major revolutionary changes in the business model may not be believed by employees or necessarily understood even if managers find it completely logical. When working at a strategic level it is easy to forget the first sentence of this chapter. Delivery of the new model will require the creation of new working relationships – these relationships determine efficiency and effectiveness of the 'desired future state organisation' (Kurogane, 1993). As a result, the change process will be different for each organisation and one formula cannot suit all businesses and even lean businesses will have followed different change processes to achieve this status. The top-down role is therefore to design and lead this transformation. Senior managers also need to find the leverage that will help individuals to overcome their natural preference for the status quo rather than to seek out a new challenge and crucially be critical of themselves and their current role.

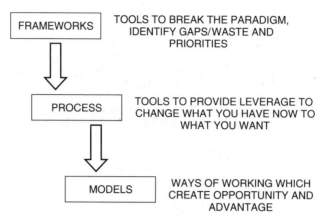

Figure 3.1 Tools to deliver real change

Research into how individuals learn to use new techniques provides some clues as to what is required (Cummins and Townsend, 1999). Two sets of teams were set similar tasks, one that had classroom training in problem-solving and a second that did not. The results suggested that a significant factor influencing team success was not training but confidence. Factors identified by the study that impact confidence were:

- Past experience of the same or similar problems;
- The environment and its approach to risk.

That is not to say that training is not important; a confident person with appropriate training will achieve more than without. It is application that is most important, and that where individuals have limited experience of doing so, support is needed. In addition, confidence depends on the level of trust in management. Figure 3.2 sets out a proven High Performance Teamwork development framework (Buffin Learning). This builds on the basic teamwork development framework introduced in Chapter 1. The right-hand side of the model sets out the changing management role required to support the progress of the team towards real empowerment. It also sets out how this is built on an understanding by managers of individual behaviours.

The model provides a general framework linking the process of change to progressively refined models of operation and provides an insight into how continuous improvement really does mean a life of never-ending learning and experimentation. Many organisations make the mistake of thinking that the application of frameworks alone will make a difference. Such organisations regard continuous improvement

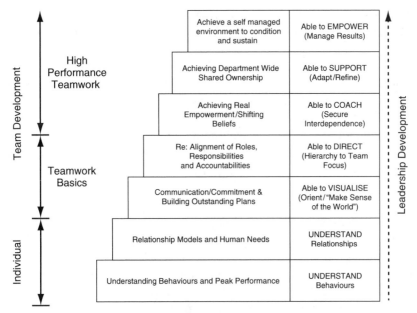

Figure 3.2 Leadership/Steps to High Team Performance

as an additional task that can be simply grafted on to existing relationships and job roles. For change 'to stick' the search for better ways of working must be central to every-one's role and this means encouraging employees to look at their roles and eliminate those elements of their job that adds low levels of value or efficient customer service. Senior managers must demonstrate this attention to detail and conditioning of workers, including making change fun rather than pain, in order to underwrite success at the business level. In practical terms the *top-down role* can be defined as:

- Setting priorities (consistency of purpose);
- Setting standards and supporting delivery (collective discipline);
- Giving recognition to reinforce the quality of individual learning and the 'right behaviour' sought of the new business design (objective feedback).

On this last point, studies into teamwork indicate that the biggest failing attributed, by teams, to team leaders is letting people get away with poor performance (Lafasto and Larson, 2002). Such a practice is effectively demonstrating apathy towards individuals who deliberately work against the new relationships needed and this cannot be tolerated. Team leaders don't always have to be liked but they must always be respected. Allowing ill-discipline at the team level creates instability in one of the key elements of the factory's value chain. It also suggests that the training of the team leader has not been totally effective and these matters cannot be treated lightly. When all is said and done, the future of the business and its longevity rest upon everyone working in the right direction. Those who run counter to this direction, after training, counselling and discipline, cannot be allowed to remain in position. They not only slow progress but serve as focal points to legitimise poor behaviour by others. The *bottom-up role* may be defined as:

- The consistent application of best practice (capability);
- The sharing of lessons learned and reflection upon success and failure (openness and learning together);
- Involvement of all stakeholders in the factory area (including trade union representatives, safety representatives and support staff);
- The problem ownership/continuous improvement (aligned goals).

The delivery of these roles is dependent on the *joint development* of:

- A clear compelling future model;
- A practical and enjoyable change process to deliver the new model that is introduced to allow operators to understand each phase of change and see it as a simple extension of what has already been achieved;
- The total immersion by management and shopfloor to establish new, more productive working relationships.

A good way of measuring the progress of this partnership is to listen to the language used. Do people talk about 'working together to make a difference' or do they blame each other? Are people at meetings positive and proactive or are they defensive and careful to avoid taking actions? Do teams make plans together or plot against each other? It is also interesting to witness whether the bias is towards highlighting past failures and sporadic events. These indicators reveal the extent of promotion and nurturing and the level of involvement of senior management and line management in engaging the workforce in change.

Creating the difference is a function of leadership. The job of every manager is to engage his team by creating a win/win partnership to promote effective and exciting change. Identifying this winning formula is not difficult and calls for listening as well as communicating. It is easy to be fooled by the belief that problems are unique to the business or process. In reality few problems are truly unique, if nothing else, a common thread is that they involve people. If you are cynical, visit a World Class organisation, talk to the people and watch how they react to problems. Close your eyes and imagine for a moment what could be achieved with this outlook in place and working across your organisation. That is what a 'World Leading Organisation' feels like. Most managers think that this can be achieved as a step change. In reality, it is the accumulation of small wins that yield the overall system needed to squeeze every waste to the zero position. Not an easy task but well worth the entry price.

3.2 Changing the business model

Jack Welch at the head of General Electric (GE) built a strong business on the model of being number one or two in their chosen markets. This led to divestment of business/acquisitions where GE companies could not achieve this through organic growth. Initially successful, the effectiveness of the model was eventually reduced as managers began to define their markets in terms that meant that they could be 'number one' or 'number two'. When the rate of growth slowed, the root cause was recognised after much soul searching and a new business model was developed. Markets were redefined in terms of current turnover being one-tenth of that market. Once the initial model had served its purpose, it was time to move on. With hindsight this outcome may seem obvious but it takes true leadership to move away from a model that has provided success in the past. No approach can be successful forever, therefore such change is inevitable. The challenge is recognising the need early enough and taking action whilst there is a choice of futures.

In practical terms, this means being driven by a passion for business growth. It is growth that ensures job security, business viability and fighting spirit – creating this vision and a sense of urgency rather than pain. It also means embracing the concept of close cross-functional rapport to secure versatility.

Figure 3.3 repeated from Chapter 1, sets out the frequently repeated business turnaround journey. When the business model being used no longer provides the winning hand in the market place, it becomes a liability.

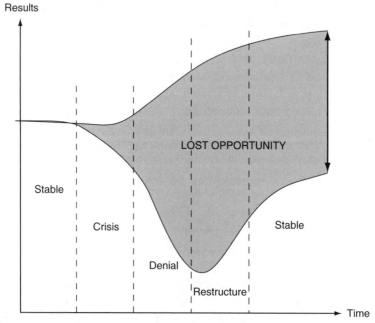

Figure 3.3 Leadership – hidden losses

Imagine that you run an organisation that supplies products to a major automotive manufacturer or food retailer. How would your business cope with the loss of such a contract? Imagine the same business dealing with a significant seasonal fluctuation, caused by exceptional weather. What pressure will that put on the organisation? How quickly would it rise to the challenge? Will departments work together or will they protect their turf?

The role of management is to create a business model that is versatile enough to cope with such likely pressures and provide direction. If a manager thinks only of today, they reduce their role to that of a progress chaser/policeman. There are few cases of progress chasers leading successful change programmes. A measure of their leadership is the level of alignment of personal goals such that the organisation moves as one to stay ahead of the competition.

The scope of this task extends beyond the shopfloor to include all elements of the value chain as illustrated in Figure 3.4. The hexagons represent interlocked and parallel management activities. The arrows indicate the deployment cascade from strategic intent to practical reality. Every programme starts therefore with a robust business strategy that is based upon delivering customer value now and in the

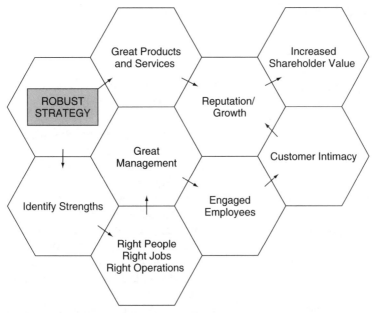

Figure 3.4 The impact of strategy on value generation

future (Womack and Jones, 1996). The strategy sets direction and the passage through the hexagons will help focus management thinking and promote consensus. No single manager can achieve these goals by themselves. The process of going through the hexagons should, as a management team, create a common sense of purpose and a common 'hymn sheet' for the management group. This united and agreed future perspective at each stage of the value chain, is fundamental to establishing a new culture. The resulting clarity of vision will help management to reinforce the versatile cross-functional business model as part of their 'day job'. The outcome is consistent on the job coaching by managers which confirms the commitment of management to integrate personal agendas behind a common vision. It is the only effective countermeasure to functional thinking (ignoring the business and concentrating upon the individual departmental responsibilities) and creation of multiple initiatives which confuse the change agenda.

Table 3.1 explores the various elements of the model and sets out the scope of the change agenda. If these are not defined proactively by management, they will be 'defined' by the organisation or perhaps its competitors.

Table 3.1 Integrating the change agenda

Component	Questions
Robust strategy	How to remain successful despite predictable shifts and external pressures.
Identify strengths and weaknesses	What do we need, what have we got and what holds us back or is a deficiency with the current state organisation.
Right people, right jobs, right operation	How to build on strengths and develop the capability we need.
Great management	How to build trust through discipline, consistency and recognition.
Engaged employees	How to create the environment where employees are fulfilled by delivery of business results.
Customer intimacy	What would we need to achieve to lead the customers' agenda.
Great products and services	What innovations do we need to lock in profitable customers.
Reputation and growth	How can we become the customers' first choice and attract the best people.
Increased shareholder value	What do we need to do to become the one to invest in.

3.3 The senior manager role

The Lean TPM treasure map provides a framework to identify areas of hidden loss/waste across the supply chain from supplier to customer. This also positions these 'gaps' against the three main policy deployment levels within the organisation. As such it defines accountabilities, provides a target for improvement resources and supports the development of a single change agenda across all functions. The treasure map logic supports the primary lean rule of understanding what customers value and what they are happy to pay for. The process will, at each stage, generate a number of questions, providing a good general learning process for managers undertaking the strategy review. Business managers engaged in this process will also find that their initial assumptions, such as the strengths and weaknesses of the current state organisation, will at later stages be tested. This deliberate and iterative process helps managers to refine plans and learn how to view the organisation as an interdependent set of commercial relationships (Senge, 1993). The strategy review process below passes a number of stages. The outputs from each should be documented so that earlier assumptions can be reviewed when during the later stages, choices become clearer:

1 Issue mapping: what are the issues facing the business (internal and external).
2 Agree the core focus for business, what is the reason for current and future success.
3 Develop the draft 3–5 year vision. Make it one which everyone can get behind. This may require analysis of options against the core focus.
4 Scenario planning. What flexibility do we need the concept to cope with (best case/worst case/most likely). Consider the response to the four main business risk areas of market, cashflow, operations and throughput volume.
5 Agree the recipe for success (include cultural development) and what are the three or four 'must do' ingredients. Identify what is at the heart of future success and the linkages between the 'must do' activities. Use the Lean TPM treasure map to identify gaps and priorities for improvement at each level of the organisation.
6 Evaluate options for delivering the 'must do' activities. Use the road map set out in Chapter 2 to identify bite-sized steps.
7 Design the change process, identify the change team (see Chapter 4), their accountabilities (see below) and supporting development programme.

Conducting these analyses as a group of cross-functional managers may take time but will result in a holistic and better-grounded change model (Kurogane, 1993). We will now explore the major wastes to be considered during these management-level discussions and research.

Wastes in the supply chain

Internal analyses and redesign of the production system is only one part of a much bigger and 'waste laden' supply chain. Most businesses, whether they care to admit it or not, are dependent upon the performance of their suppliers. For decades, supply chains have been enshrouded with mistrust and adversarial relationships (Cox, 1996). The approach to suppliers has, therefore, been to exclude them from anything beyond selling their product and quoting a price. However, the supply chain is also founded upon relationships. Suppliers provide products or services which are not core to the business or which can be made by them more cheaply. For modern firms, traditional supply chain management practices will not support the creation of 'manufacturing-led' competitive advantage and require a similar approach to close working relationships if these important external resources are to be exploited for advantage. Improving interfaces between the organisation and its suppliers affords great opportunities to yield savings in a relatively short space of time. In some cases these can dwarf the savings that could be made by focusing on the internal value stream.

It is strange to think that, for decades, adversarial relationships and deliberate exclusion of suppliers from involvement with the customer has been the main form of relationship for most industrial companies. It has been a default strategy that has not served industry well; instead resulting in high minimum order quantities, vast amounts of time spent tendering for suppliers, buying from sources in cheap labour economies (but ignoring the true cost of supply) and a lack of synchronisation between forecasts and the actual manufacturing of products. The next sections will explore some of these issues.

Understanding how to exploit and eliminate these wastes can have a tremendous impact on the efficiency of material flow as well as 'open the door' to inter-company collaboration on such issues as the design of new products, stock and cost reductions. A good supply chain does not happen by chance, it is designed. Senior management must therefore focus on the design and refinement of the value stream of suppliers, and here it will be found that historic relationships have resulted in six supply chain wastes and practical areas where the value generating capability can be improved. These wastes affect the availability, performance and quality of the materials exchanged. The elimination of these wastes enhances material flow through the external value/supply chain and embodies two of the five main principles of 'lean thinking'.

At a supply chain level, losses are considered under three main headings (Table 3.2):

- Logistics concerns the way in which the supply chain is linked and choreographed. This includes planning rules/forecasting and flow management/flexibility of response;
- Transformation concerns how well value is added covering the effectiveness of the new products/technology design/specification as well as the optimisation of current technology;

Table 3.2 Assessing supply chain wastes

		Logistics	Transformation	Customer value
Senior Manage-ment. Supply Chain (S2C)	Unplanned	Planning rules	Design effective-ness	Unmet customer needs
	Planned	Flow manage-ment	Non-value adding activities • Transport/ Motion waste	Late or incomplete delivery Supplied defects

Goal: Effective use of potential resources (time-based competition)

• Customer value concerns how well value is created and protected including current customers and the ability to satisfy unmet customer needs.

Planning rules

Frequent changes to the production plan indicate that the supply chain is not in synchronisation with customer demand. The more frequent the changes the higher the risk of customer service failure. In most cases, planning rules are never questioned but often include untested assumptions concerning capability to supply. It is not unusual for a plan to be developed in detail well in advance only to be consistently ignored in the 'heat of battle'. In addition to internal problems and risk to supply these unpredictable swings in order volumes mean that suppliers resort to inventory buffers or engage in expediting work to protect their manufacturing process (Figure 3.5).

For most supply chains this demand amplification (between what is forecasted and what is actually taken) can be huge resulting in stockpiling by all companies. These stocks add cost and slow inventory turns no matter how good your door-to-door OEE figures. The level of initial plan achievement is a good indication of how well planning parameters co-ordinate resources in line with true demand and how well the customer business has been in levelling demand to within

Loss Category	What is it	Why is it bad	How can we find it	How can we reduce it
Planning rules (S&OP process)	Late changes in production schedules, inappropriate planning assumptions or poor forecast error levels	A poorly tuned supply chain can result in fire fighting to assure stock availability as well as additional costs, wasted effort and "just in case" behaviours	Monitor unplanned events across the supply network, Review S&OP planning assumptions against actual performance	Demand management analysis of demand patterns. Improved information flow. Delegate detailed planning decisions
Flow management	Under utilised assets, slow moving inventory or excess distribution costs	Spare capacity and surplus stockholding adds to fixed costs, and masks shifts in customer demand patterns	Macro map supply chain to identify value added time, Portfolio review, Stockturn against ABC demand profile	Improve design for manufacture, Mfg flexibility. Target non value adding processes
Design effectiveness	Technology which is difficult to use, look after or is wasteful in terms of resources	Poor design effectiveness leads to higher transformation costs and increases the level of intervention, skill levels and risk of defects	Analysis of energy, tooling costs, criticality analysis of design, Day in the life simulation	Early Management of products and capital projects, targeted capital projects
Process optimisation	Inconsistent process quality or resource needs	Lack of control of chronic losses results in higher levels of inventory and quality defects	Stable or declining improvement trends	Use Quality Maintenance to establish and deliver optimum conditions
Unmet needs	Ability to develop and launch successful new products and services	The ability to create new value is essential to business growth	Slow growth in profits from new products, stable or declining time to market	Excitement feature VOC innovation, Product life cycle management
Poor customer service (reputation)	Lost sales, weak customer loyalty or reducing market share	Indicates lack of understanding of market shifts and potential loss of competitiveness	Share of Market and Customer business, VOC survey	Total cost of ownership analysis. Basic and Performance VOC innovation

Figure 3.5 Dealing with supply chain wastes

controlled tolerances with the supplier. A plan should be realistic and achievable; that is realistic in terms of meeting customer expectations and achievable in terms of time and resources available. Poor planning parameters can lead to significant hidden waste throughout the organisation as people engage in expediting or sorting out quality defects before they affect the internal OEE calculations. Reducing the chaos of poor planning management has many benefits, not least in ensuring that processes are available to produce what the customer wants in the smallest batch sizes possible (maintain a low production cost). Demand amplification and planning chaos also mean product costing information is based on parameters which do not reflect reality as production consumes overtime, batches are cut short and all manner of intervention is needed just to get products out of the door.

Demand amplification also means, although improvements are made at the door-to-door and equipment levels, this potential is not translated into real benefits. Until improvement activities reduce inventory little has been achieved. Effective businesses use these improvements to reduce inventory and to lower costs of operations (Standard and Davis, 1999). Staggering improvements in changeover time which do not reduce total inventories deliver little real business benefit. This important point also has implications for generations of engineers who have sourced the quickest and fastest processing machinery when given a capital expenditure budget to replace existing machinery. Buying the 'state of the art' machinery, to replace existing assets, must result in a redefinition of the stocks needed to support the business. If this does not happen then the potential benefit of the quick cycle time of the new machine is lost in a mass of inventory. In reality, value added of factory operations has worsened because excess inventory has not been removed.

This form of investigation also applies to other planning parameters. For example, an investigation into delivery failures in a chemical plant revealed how planning rules based on large batch sizes meant that despite spare capacity, opportunist orders won by salesmen were impacting on orders to customers who had given long lead times.

Flow management

Often slow moving or specialist products are a variant of another product, possibly a foreign language version or pack size. Storing intermediate stock or assemblies and making such products to order can provide the flexibility to deal with such difficult to forecast demand items. In this example, the intermediate stock is the decoupling point between supply and demand; the point where resources are focused to meet demand. This is the domain of flow management. Lean TPM seeks to move the supply decoupling point as far back towards the raw material supply as possible. If demand can be met by making to order within the customer order lead time this will provide the lowest cost supply option.

Typical Flow Management losses include spare capacity and excess inventory.

Spare capacity is not necessarily a bad thing where it provides flexibility to protect against seasonal or customer distress purchases. These events are often predictable and can be controlled for many companies. Even businesses affected by the weather temperature or sunshine tend to have product characteristics (shelf life) that allow these events to be accommodated without causing chaotic production. The ability to flex capacity can, even under these conditions, be an important component in maintaining customer loyalty and providing customer satisfaction and one of the predictable gains from TPM is increased capacity. The following and the simple case study below shows a typical improvement curve over a 5 year period (Figure 3.6). The company identified that meeting current planning standards, improving set ups and reducing scrap is the route to doubling OEE and with it capacity that can be sold as a growth strategy to increase the numbers of customers with whom to trade or to introduce new products. The benefit of delivering the forecast business growth at the improved OEE level included profit margins increased by over 200 per cent. Had the increased capacity not been required the gain would have been around 40 per cent. Still worth having but clearly a lost opportunity.

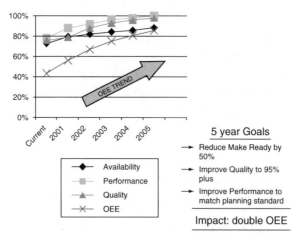

Figure 3.6 Understanding the future impact of OEE (a case study)

In defining a change mandate, by looking at capacity issues and the predicted benefits of a Lean TPM approach, the ability to sell manufacturing capacity is important. The most effective way of engaging the workforce behind the improvement agenda is to follow a growth strategy. For the workforce, growth is a way of avoiding problems with what to do with labour displaced by improvement activity. Without this approach, improvement activities will falter as individual workers correlate improvements with the loss of jobs for co-workers. The win-win proposition is to achieve a growth rate which is equal to or better than the rate of improvement. In most industries the biggest single factor influencing profitability is order volume so this also makes commercial sense.

Surplus inventory is often a result of trying to plan too far ahead and in too much detail (Womack and Jones, 1996). In particular:

- The longer the planning time-scale the more difficult it is to predict what will happen;
- The more detailed the plan, the less willing people are to change it to reflect reality.

To lean businesses, inventory is classified as an 'evil cost' – it is sleeping money that is not earning a return. As such, lean businesses employ strictly controlled kanban stocks to limit this exposure to idle money and pay attention to maintaining flow. Also, asset care activities are used by lean businesses to ensure assets don't fail and require safety stocks to cover for this failure. These stocks are in effect an apology for poor maintenance. Low inventories make sense both in terms of 'business logic' and reinforcing the importance of asset care and development of diagnostic skills. The other reason for deliberately lowering safety stocks is to force improvement and allow abnormalities to be detected quicker and action taken by the appropriate operational personnel.

The planning process does not have to be complex. There are only two planning parameters to manage:

- How much to produce;
- When to produce.

Minimum co-ordination is achieved when:

- Medium-term planning monitors demand trends to assure that outline resource/capacity allocation rules can match true demand;
- Detailed planning is delegated as far down the organisation as possible.

In this way, planning decisions can be made as late as possible rather than attempting to anticipate in advance day-to-day conditions such as labour availability or last minute changes in customer demand.

Experience shows that in some industries it is possible to use simple rules to co-ordinate up to 80 per cent of the production volumes. For most businesses demand profiles can be broken into three categories of product manufactured at the business:

1 Runners (5 per cent of products providing 50 per cent of revenues).
2 Repeaters (15 per cent of products providing 30 per cent of revenues).
3 Strangers (80 per cent of products providing 20 per cent of revenues).

Demand for products classified as runners is fairly predictable and finished goods' safety stock levels can be low. The decision on 'how many' or 'when to make' can be controlled by simple 'two bin system' or kanban where work is pulled by actual demand. The variation in demand for repeaters is higher so that it is only possible to predict over-all demand. These are the items where some batching of production will be necessary. They can be made at any time capacity is available.

The variation of month to month individual stranger demand is very high making it impossible to forecast accurately. It is generally possible to forecast demand for groups of stranger items and from this how much capacity they may require. So although these are the items that can only be produced to order, it is possible to reserve capacity for them keeping lead times to a minimum. If stranger levels are lower than forecast, the capacity can be reallocated to make runner or repeater products.

The net result is that flow management tactics should reflect the predictable patterns of demand profile. Ideally, these should leave the choice of what to manufacture as late as possible by using simple capacity allocation rules which can be managed as close to the point of manufacture as possible. Under these conditions, opportunities to use available flexibility will be encouraged as will activities to further increase flexibility to changes in demand.

Transformation

Process design: To deliver high flow, low defect, low cost products, processes need to be intrinsically safe, reliable, easy to operate and maintain. It is common to find that the opposite is true. Equipment is difficult to operate and maintain.

Another design waste is that of over-specification of equipment. This has a number of costs associated with it, not least that of increasing the amount of losses during processing. For example, an oven that is bigger than needed will take longer to heat up, cost more to run/maintain and usually cost more to buy. In addition, functionality which is not required adds to the complexity of the design making problem-solving and optimisation more difficult. It is also important to identify where potentially useful features are not used because they are in a poor state of repair. These issues go right to the heart of the 'automation debate' and the 'right sizing' of productive assets to do the job expected of them using capable technology. All too often, neglect and even engineering experimentation has resulted in an asset base that is in need of review. It is interesting to note that many 'world class' organisations have basic policies which involve precluding experimental and unproven technology from the workplace as these assets are, by nature, problematic and risky. Also many of the 'world class' organisations, such as Toyota, also don't automate a process that has not been perfected by rounds of problem-solving and engineering assistance (Monden, 1983).

With the relentless pressure from customers for new products, the ability to deliver new products and equipment that achieve their full potential is an important strength. This will therefore have a major impact on the growth strategy of the business.

It is said that if you want to know how your prospective partner will age, look to the parents. If you want to predict how well your future technology will support competitive advantage, review the current technology to provide:

- An assessment of current technology/product platforms and their fitness for purpose. This should highlight strengths, weaknesses and where current functionality is not being used.
- An indication of the effectiveness of previous decisions making processes to support learning from experience (compare this with the Early Management process described in Chapter 6).
- A guide for the setting of future technology standards (see Table 3.3).
- Clarification of the input required from Commercial, R&D and Operations influence in delivering a better process.

Optimisation and reducing non-value-adding activities

The Lean TPM journey passes through the stages of zero breakdowns towards zero defects. On the way, as organisational capability improves, non-value-adding activities are removed. This is a key

Table 3.3 Assessment standards for technology design

Score	Commercial	Operations	Technical
1	A serious risk to the basic business model	Extremely difficult task	A serious risk to process control
3	Provides competitive capability and advantage	Able to avoid breakdowns due to processes of asset care and inspection	Acceptable process controls and achievement of tolerances
5	Provides a market winning advantage	Easy to do the process right and foolproof	Process is easy to optimise with 'no touch' operations

enabler of progress towards achievement of market/customer leading capability.

Transportation and motion losses are common areas of focus for removal of non-value-adding activities. Customers are not willing to pay for the fact you move products over 2 miles within your factory before it is packed and ready for shipment. Travel distances increase the likelihood of damage and 'world class' organisations seek ways of lowering travel distances to achieve much better just-in-time performance. Even process plants engaged in Lean TPM are working to the concept of 'pipe-less' plants where buffers between processes are minimised. Pipes are a hidden form of transportation that have their own difficulties including the phenomenon of 'hammer rash' which is the technique used to unblock pipes by hitting them with rubber mallets to increase flow. 'Pipe-less' factory designs, for process plants, makes it possible to produce smaller quantities economically and to reduce capital costs, plant footprint, new plant delivery lead times and improve return on investment. This issue, especially making visible this form of waste, is central to the technique of value stream mapping which will be explored and covered in more detail in Chapter 5.

The solution to these problems involves a new relationship and alliance between the marketing department, maintenance, operations and R&D functions to deliver an asset base and production system that is sized to deliver what customers value. This is one of the fundamental drivers of Early Management (see Chapter 6) and Quality Function

Deployment – shortened to QFD (Cohen, 1995). QFD, incrementally decomposes the features highlighted by the Voice of the Customer profiling into actual specifications and operating conditions for the asset to produce these values (Bicheno, 2002). A simplified but powerful example of QFD analysis is included at Figure 6.10.

Customer value

Unmet customer needs refers to consumer products like Sony Walkman or personal computers as well as business innovations like outsourcing and line side deliveries. Once unheard of, now a part of the fabric of life. Research (Christensen, 2002) suggests that targeting unmet customer needs resulting in disruptive innovation, produces products and services with a one in three chance of success compared to a one in fifteen chance of success for 'me too' products. The research also highlighted that when leading companies were toppled, it was almost always when disruptive technologies emerged. Although this research focused on technology-based products, the findings are relevant to many other industries.

Understanding the 'voice of the customer' is, therefore, an essential capability for any manufacturing business. Failing to understand the purpose for which the product is being purchased and believing that your products can remain unchanged for many years is no longer credible. The car industry is a good example. About two decades ago, the materials and components used would have been relatively standard and unchanged. Today, the modern car industry upgrades these items every 6 months to allow improvements to be assembled into the vehicle. Even materials such as steel have been subject to improvements in bodywork warranty and anti-corrosion performance. In short, it is easy to miss and not meet customer needs.

The importance of the service/product package should not be underestimated. In retailing one of the areas of increased margin in recent years has been the offer of extended warrantees. Finance deals are an increasing source of revenue for automotive manufacturers. In business to business industries, understanding the customers total cost of ownership (TCO) analysis can provide ideas for reducing costs to the customer. This also has the potential to increase supplier revenues because the product/service offering is more valuable to the customer. It may even result in different ways of charging for the product/service. For example, a supplier of oil and gas platforms undertook a project to build a platform for a major customer where part of their revenues were paid based on the barrels per day output from the platform. They were able

Table 3.4 Understanding customer needs and the impact of technology

		Technology	
		Current	New
Customer needs	Met	Replacement products	Disruptive products
	Unmet	Incremental products	Customer leading products

to share the gains with the operator of improving the design in ways which reduced capital cost and increased output.

Review the past performance of your business against the four categories set out in Table 3.4: what has been tried, what was successful, where is the future potential. Further analyses should consider *'What would a competitor need to do to win business from this company?'* This analysis, at least, provides clues as to where to investigate to create a portfolio of product and service offerings to satisfy the modern customer.

These analyses typically illustrate how collaboration within your supply chain provides an effective countermeasure to a stagnant marketing mix of products and opens new opportunities to engage suppliers with the customer market for mutual gains.

Customer service targeting of 'On time in full' is in theory the goal of every supply chain yet many supply chain strategies accept a trade-off between cost and service levels. Increased reliability and flow potential will make it possible to reduce lead times and improve forecast accuracy thereby eliminating unnecessary stocks. This again has an impact on customer loyalty. Research into 'why customers change their loyalties to a certain manufacturer?' suggests that around 10 to 15 per cent of customers change suppliers due to changes in circumstances and around 30 to 60 per cent routinely assess supplier performance. Understanding the losses associated with delivery performance is therefore important if senior managers are to understand what drives customer decisions so that they can drive supply chain improvement priorities.

3.4 The middle/first line manager role

In any change programme, it is the middle management level that dictates the rate of change and is also the most problematic to deal with. This is the level of the organisation where the strategic intent meets the white heat of shopfloor reality. Often unintentionally this layer of management can hinder change by translating top-down input by filtering out what is seen as impractical.

The same can happen when translating bottom-up input to remove unrealistic requests that will not be considered. This is also the level of the business within which politics and 'guerrilla warfare' tactics have traditionally been used in power struggles between departments (Rich, 1999). Of note is the power struggles between operations and marketing (Brown, 1996). This is ironic in that both these departments rely upon each other but more often than not this tension still exists in modern manufacturing firms. The problem between these two departments rests upon the issue of 'customer service'.

For marketing, the operations are a drawback and constraint that prevents the business from moving forward. For operations staff, 'marketing personnel' are often regarded as 'know nothing' salesmen, who have little appreciation for the engineering input, capacity management and other issues required to deliver products against the contractual requirements set by the sales and marketing staff themselves.

This relationship, however, lies at the crux of 'world class' manufacturers. Such organisations have managed to create a superb brand for the quality of product supplied and to have correctly and effectively harnessed the manufacturing operations of the firm as a means of competitive advantage. They have achieved this through breaking down the internal boundaries and improving communications to achieve a mutual respect and understanding across functional boundaries.

Door-to-door losses

The improvement focus of first line managers (FLM) are door-to-door losses (Table 3.5). These loss categories highlight problems due to gaps in top-down policy and difficulties in shopfloor operation. As such they provide a framework for managing upwards as well as one for coaching FLMs to develop a wider appreciation of the business.

Door-to-door losses are categorised under the headings of:

- Preparation;
- Co-ordination;
- Adherence.

These forms of loss represent targets for improvement and gaps in the fabric of management. They help to unpick the complex cause and effect relationships which result in lower true customer service and business

Table 3.5 Door-to-door losses

		Preparation losses	Co-ordination losses	Adherence losses
First Line Manager (D2D)	Unplanned	Supply failure	Co-ordination losses	[a]Schedule adherence
	Planned	Ancillary operations including planned waiting time	[a]Processing losses	Material waste (zero goal)
		Goal: Effective use of internal resources (Lean Enterprise)		

[a] Part of the seven classic Toyota wastes.

effectiveness. The outputs from this analysis helps to direct resources towards the most important improvement potential (Figure 3.7).

Preparation losses

Preparation concerns those tasks carried out to support the production process. This includes inbound logistics and administration as well as maintenance and cleaning activities. Often such activities are ignored despite their potential for improvement.

Waiting time can occur when resources are not available to begin production. The reasons can be many and various. Some of the most common reasons include inbound logistics problems, poor shift control and poor utilisation of engineers, even inconsistent or irregular production scheduling. There are many reasons why the schedule itself could be irregular and exploration of these issues often finds many problems associated with 'dead data' in the planning system. Other issues include not scheduling resources because of unexpected interventions made by the sales department to reprioritise orders and accommodate sales of products within the stated lead time of the company. These issues will also be reflected in low schedule adherence. The solution involves the planning process and focus on creating stability for manufacturing in terms of getting materials available and communicating manufacturing schedules such that labour is available.

Ancillary operations includes operators placed on the line to watch for falling products or sortation (and assembly) line processes where the level of work is not balanced. These 'non-jobs' become accepted

	Category	What is it	Why is it bad	How can we find it	How can we reduce it
Preparation	Waiting	Resources materials or instructions fail to reach the first process as planned	Can result in wasted effort and increase the risk of customer service failure	Monitor delays to campaign start times. Analysis of inbound logistics	Improve information flows, Target total cost of supply, Supplier managed stocks
Preparation	Ancillary Operations	Additional activities needed to prepare for the next campaign, includes cleaning, maintenance or pre production inspection	These resources can be hidden in fixed costs and not subject to CI. Their impact on productivity can also be undervalued leading to under investment	Value mapping of door-to-door internal supply chain. Losses. Day in the life simulation, Analysis of fixed costs	Process reengineering to eliminate, combine, simplify ancillary operations
Co-ordination losses	Unplanned Intervention	Unplanned delays in feeder to fed processes or additional activities to expedite production	Interrupted process flow means higher level of intervention, co-ordination and space requirements raising fixed costs	Line studies, value mapping, labour productivity KPI's, WIP levels, Resource usage	Synchronous production techniques e.g. kanban, Improve process reliability
Co-ordination losses	Non Value Added activities	Activities such as transportation, double handing or in process inspection delays	Extends production lead times, adds to WIP and intervention levels. Can result in the need to expedite backlogs	Value mapping	Improve product flow using feeder to fed processes and pipeless process plants
Adherence	Schedule adherence	Inability to produce and deliver as planned. Includes overproduction	Results in plans with inbuilt buffers to allow for "unforeseen" but avoidable problems	Areas requiring routine expediting, low correlation between planning standards and actual time taken	Review basis of planning standards, simplify scheduling and delegate short term planning activities
Adherence	Total Cost lost	Accepted resource waste/ material give away both formal and informal	Can hide wasteful practices or potential for improvement	Target total yield rather than performance vs cost standard	Product and tooling design. Process optimisation

Figure 3.7 Dealing with door-to-door losses

elements of the work and often involve operators fetching chairs to sit on while they watch the machines for which they are responsible. Such waiting can be reduced through multi-process operations using walking operators who tend to many machines and this is a well-known technique employed by lean organisations. With process optimisation, the need for such watching and interventions to cover for material flow failures can be eliminated. Such tasks are boring and often dangerous. Their removal improves safety and quality as well as productivity. Other examples are maintenance, product testing and cleaning. These are essential activities but like set up and adjustment time, they should be subject to review to reduce the time taken whilst maintaining quality standard. Some carry out such tasks at weekends or on night shift and, although this might seem a reasonable tactic, it is often expensive (adding 'unsociable shift premiums', removing the opportunity for learning and problem-solving between teams).

The solutions to these problems include the need to re-industrial engineer processes, to understand the costs associated with the losses and also to engage problem-solving activities with the various factory teams to highlight these wastes in an attempt to eliminate or reduce them.

Co-ordination losses

In many industries, equipment is not the bottleneck resource, it is people. Here the availability of labour or specific skills can be the difference between profit and loss. Reducing and eliminating co-ordination needs to evolve self-managed systems is the equivalent of equipment debottle-necking. Part of this is the removal of non-value-adding activities and sources of unplanned interventions.

Ideally co-ordination within the value generating process should be automatic and aided by high levels of visual management. The aim is to develop autonomous or self-managed systems.

There are many potentially self-managed internal co-ordination approaches including:

- A total pull system (where product variety allows);
- Only scheduling the real bottlenecks in the factory;
- The use of finished goods stock levels as the trigger to launch materials to the manufacturing system.

The need for management intervention prior to deciding what to make next or waiting for sign off by Quality Control (QC) is an indication of poor job or workflow design and a lack of process reviews to reduce or eliminate these sources of failures. This applies equally to tasks such as maintenance, production or project approval and again a re-engineering approach, using cross-functional and operations teams, is needed to combat these issues.

Adherence

If door-to-door activities are under control, they should lead to easy adherence to planning rules and zero waste. The two main loss areas concern adherence to time plans and adherence to material and resource costs. For this activity, material and resource waste are reviewed against a zero base to scrutinise planned or standard waste allowances such as material waste. Typically, material costs are three times that of labour and should be subject to a corresponding level of focus.

Schedule adherence concerns the detailed schedule covering the internal supplier/customer work flows as well as the delivery of end customer orders. This is a major loss area for most manufacturing businesses and is the source of many business failures. Instability in the basic schedule (the information that triggers production) is costly and causes products to be launched too early, in the wrong sequence or too late for the needs of the customer. Furthermore, when order levels are volatile and the schedule unstable, some will 'best guess' the schedule and inflate requirements so that orders to suppliers bear little resemblance to what is being sold. The inevitable outcome is that the business has stocks of what is not selling and no stocks of what is.

Planned wastage covers the waste that is 'allowed for' in the product costing. This can include standard material waste, effluent disposal, cutting fluids, packaging, tooling, water usage and other high cost consumables that are traditionally considered 'overheads' but should be treated as direct costs. This is worthy of a major review by a multi-disciplinary management team including sales, marketing, planning and operations management. This is particularly useful where batch sizes are variable and, on certain jobs, the company could be losing money. This information is also relevant to the sales and marketing department who, armed with this knowledge may be able to negotiate a better sales price for the difficult/costly to make products. Cost deployment is discussed in more detail in Chapter 6.

3.5 Calculating the overall equipment effectiveness

The traditional hidden loss focus in a classic TPM environment is at the equipment level. Here, Overall Equipment Effectiveness (OEE) is the key performance indicator. The OEE has three components and is calculated as follows (Table 3.6):

Table 3.6 The OEE calculation

Availability %	Performance %	Quality %	
$\dfrac{\text{Actual run time}}{\text{Planned run time}}$ $\times$	$\dfrac{\text{Quantity produced}}{\text{Theoretical quantity produced}}$ $\times$	$\dfrac{\text{Quantity produced right first time}}{\text{Quantity produced}}$	$=$ OEE%

The simplest way to explain the advantages of the measure is to use an example assuming the production statistics in Table 3.7:

Table 3.7 Exploring the OEE calculation

	Description	
A	Planned run time	20 hours
B	Set up	1 hour
C	Breakdowns	1 hour
D	Actual run time is A − (B + C)	18 hours
E	Theoretical rate per hour	200
F	Theoretical production when running E × D	4000 units
G	Actual production from production sheet	3000 units
H	Quality rework	100
I	Right first time production G−H	2900

The OEE calculation would be as follows:

$$\frac{18 \text{ hrs}}{20 \text{ hrs}}\% \times \frac{3000}{3600}\% \times \frac{2900}{3000}\% = \text{OEE}\%$$

This equates to an OEE of 73 per cent (as shown below).

$$90\% \times 83.3\% \times 96.7\% = 73\%$$

Even though the individual components of the calculation are reasonably high, the OEE itself shows plenty of room for improvement but where to start? Naturally we want the process to be available when we want it, to run a maximum performance rate and produce 100 per cent right first time quality. Each component of the OEE is linked to two areas of loss. Each loss is a different sort of problem so by identifying the priority loss area, you also identify the techniques to reduce it. A further sophistication of the approach is the use of the best targeting. If the OEE is calculated over 3 or more weeks, the individual results will fluctuate (Table 3.8).

Table 3.8 OEE analysis

Week	Availability %	× Performance %	× Quality %	= OEE%
1	90%	83.3%	96.7%	73%
2	85%	85.3%	96.6%	70%
3	95%	81.6%	96.8%	75%
Average	90%	81.6	96.7	73%
Best of the best	95%	83.4	96.8	78%
Gain	5.0%	1.9%	0.1%	5%

This gain from 73 to 78 per cent provides a realistic and achievable 1 year improvement target for the process. Not only that but the highest area of fluctuation is usually a good place to focus on. In this way the regular recording of losses provides an insight into the shopfloor reality to identify weaknesses and monitor improvement trends. The OEE measure is a highly punishing but practical way to focus improvement attention and promote the change mandate within the factory. The OEE value for an asset or linkage of assets as a whole is a great way of gaining stability by focusing improvement activities at each stage in the production process with the outcome that each percentage improvement in OEE performance increases the flow of products through the factory. The OEE figure is therefore most commercially important when applied to the bottleneck process and all operations after that asset.

Designing the OEE measurement approach

In the initial applications of TPM during pilot programmes or in jobbing shops the OEE measure is typically shown as 'individual machine effectiveness' but in many cases the measure is most useful when applied at a cell or production line level. Here it is important to design the application of OEE carefully so that it provides a reliable measure of effectiveness which does not suffer from statistical noise or fluctuation due to external factors. It is worth remembering that 'effectiveness' is a measure of how well we achieved what we planned to do.

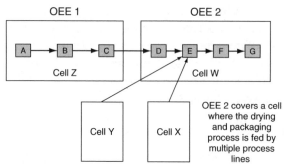

Figure 3.8 Designing the OEE measurement system

A well-designed OEE measure should support the process of learning from experience and identifying what stopped achievement of the plan and dealing with the reasons why things do not go according to plan.

Figure 3.8 illustrates a production process where cell W is routinely fed by cell Z and occasionally fed by cells X and Y. Initially, the company calculated the OEE across cells Z and W as a single flow. This resulted in fluctuations in OEE. In the weeks where cell X/Y fed into cell W sometimes more than 100 per cent performance was recorded. Creating separating measures for cell Z and cell W provided far more meaningful results.

What is important is the OEE trend, rather than absolute OEE levels. It should always be improving. It is easy to fudge OEE measures so that they look good. This is one of the reasons why it is meaningless to compare OEE results on different items of plant (such as a heat treatment furnace and a metal press). As long as the OEE information is measured on a consistent basis, this provides meaningful management information to direct and confirm progress towards waste elimination. Figure 3.9 illustrates how the OEE trend behaves as equipment progresses from an out of control to an in control condition. Over the first 6 months of the chart the OEE trend shows little overall change. Note how the availability and performance curves cross over. When availability is down, performance is up. In this case it was due to increased pressure to produce output targets following time lost due to breakdowns (a fairly common industry pattern) but, in theory the peak performance should also be possible when the availability is high. The inability to get both should therefore prompt investigation especially as the quality rate is fairly stable because of measures to assure quality. Secondly, the analysis will prompt managers to investigate why, on occasions, speed losses ensue because of quality assurance routines. In this way the OEE measure is a means of

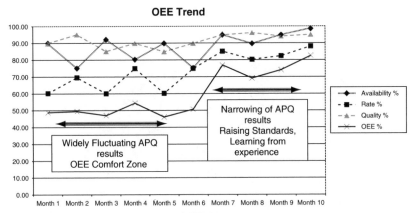

Figure 3.9 Characteristics of OEE improvement

telling us much of what is wrong as it is also a quantifiable way of seeing positive impacts of improvement activities.

In the last 4 months, as the improvement process begins to take hold, there is a noticeable narrowing of availability and performance rates. The two curves begin to move in the same direction and with less 'swing' or deviation. This indicates that problems are being dealt with and that learning from experience is taking place.

The 'door-to-door OEE' (D2D) of each cell is calculated by treating the cell as a 'black box'; here the OEE of the bottleneck will equate to the OEE of the cell. Even where the OEE is calculated on a D2D basis, it is still important to record losses incurred at each machine. Recording should include door-to-door and floor to floor loss categories. This is because on the shopfloor, the most important measures of improvement are the level of losses rather than the OEE itself. OEE measures what we have achieved, losses measure what we need to do better at. Figure 3.10 illustrates the level of losses occurring on a month by month basis. The fluctuation show that despite the OEE becoming more stable in the last 4 months, speed and quality losses are clearly not under control. As will be discussed in Chapter 5, the first priority is to get in control.

3.6 Shopfloor team

The classic six losses, monitored by the OEE measure, are well-documented and these are summarised in Table 3.9. These losses are categorised by asset availability, performance and quality issues; each of these being sub-divided under planned and unplanned loss headings.

Monthly Losses

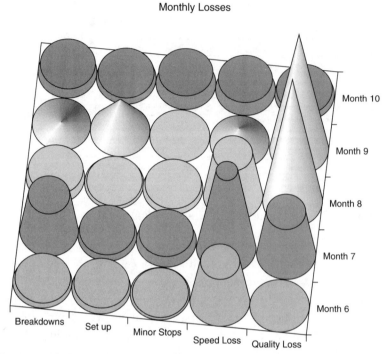

Figure 3.10 Targeting loss functions

Breakdown losses

A breakdown is a sporadic failure (such as a belt snapped or a shaft sheared) and these are different to minor asset stops. It is the issue of breakdowns that remains in the minds of operations staff although it is

Table 3.9 Equipment level losses

		Availability	*Performance*	*Quality*
Equipment (F2F)	Unplanned	Breakdowns	Idling and minor stoppages	[a]Defects and rework
	Planned	Set up and adjustment	Running at reduced speed	Start-up losses

Goal: Effective use of assets (Lean Operations)

[a] Part of the seven classic Toyota wastes.

not always the case that these are the largest losses to production when accumulated in an OEE analysis over time. Chapter 5 sets out how to achieve zero breakdowns.

Set up and adjustment

Time spent setting up the equipment is time which cannot be spent producing and so this is considered a loss. Unnecessary adjustments are also included under this heading. Examples of this include:

- Not getting the set up conditions right for the product at the beginning;
- 'Fine tuning' by operators and techniques to maintain control of the machine;
- Frequent adjustment of materials fed to the machine such as the management of reels in the production of electronic circuit boards.

Set up losses can be huge and are often ignored by operations staff who have simply grown to live with long times during machine changeover and ramp up. World leading plants achieve zero interventions or no touch production and have through sustained efforts minimised this form of downtime.

The approach to reducing losses includes an analysis of methods, by operations and technical staff, to minimise the equipment downtime using quick changeover and SMED techniques. SMED stands for Single Minute Exchange of Dies and is a procedure used to reduce the machine change over to 9 minutes or less. The ultimate aim is to achieve 'one touch changeovers' where the operator simply presses a button and the product is changed. The goal is to achieve a slick 'Formula One' motor racing 'pit stop' when carrying out a changeover. For industries such as glass making where the setting of a tool can take an hour but ramp up of glass flow can take 5 hours, this form of improvement programme offers a tremendous financial, capacity and customer service reward to the firm. From a 'door-to-door' basis, cleaning activities and material co-ordination should be subjected to the same analysis even though, in many industries, cleaning is considered a 'low grade' task and is consequentially ignored. The key to reducing cleaning times is to reduce the scattering of dust and dirt, which in turn extends the workable life of rotating equipment (lower wear) creating a double benefit for the business and material flow performance.

Idling and minor stops

The term 'minor stops' does not necessarily refer to the length of the stoppage but is a means of classifying all those stoppages that often go unnoticed (until measured) but whose cumulative effect can be enormous and drastically reduce operating time (and increase worker frustrations). At an automotive manufacturer a fire was caused by the build up of heat due to the density of spot welding tips in a particular body section; the team learned how to put the fire out quickly and restart the machine. They treated it as a minor incident despite its potential severity. At a cement plant, blockages could stop the plant for half a shift and the response was to clean out the pipes and restart. In each case the problem had not been solved, it had not gone away. This is a significant difference to a breakdown. With minor stops the first major hurdle to solving the problem is believing that it can be solved and breaking customary practice (remember the 'hammer rash' problems mentioned previously). Minor stops are highly costly to any manufacturing business; they are annoying to the operations staff and an irritant to maintainers (who are often unaware of these events). The solution is to make these problems visible by correctly documenting these failures and applying problem-solving activities or to engage devices which sound an alarm to prompt a rapid response to stoppages.

Reduced speed losses

There is often heated debate about how fast the process can run and what standard to use for performance rate calculations. Typically, however, reduced speed losses often occur as a pragmatics tactic to reduce quality defects. For companies who want to improve, there is no excuse for not applying the maximum rate possible for the process. This could be the engineering rated speed of the asset when it was bought or its highest recorded capability. This will produce a lower OEE result but the potential will be clear. These losses can also be calculated and equated into financial losses and cost of lost production. For many businesses this is one of the most difficult losses to correct. This is, as most world class companies will confirm, the stage that demands the most skills from the engineers and operator teams. This is also the loss that results in some of the most spectacular innovations which can be incorporated in the purchase specifications of the next generations of productive assets.

Start-up losses

These losses occur at the start up of production assets and this is typically the major improvement area for process plants, particularly where the process chain is fairly simple such as oil extraction. This loss is also influenced by the set up activity and in many cases, where plant is left idle for periods of time, the quality of cleaning after the close down can have a major impact on the time taken to get the asset/product back to specification.

Quality defect and rework

For most organisations quality defect reduction presents a major opportunity to improve productivity and reduce costs. The true cost of poor quality is typically huge even though most defects can be prevented using focused engineering and operator efforts. The goals of any improvement team must therefore include zero defects and a determination to waste nothing in servicing the customer. This sounds a perfectly logical objective but even after decades of promoting quality management most customers would be horrified at the extent of reworking and scrap. Such mistakes have to be paid for and the person that does so is the customer. Poor quality performance also has a vicious circle and tends to result in increasing production batch sizes, paying operators to rework products and also using too much bottleneck capacity (or using it twice) to make something that should have sailed through the production process.

However, quality losses remain one of the least understood areas of loss and involves, not the measure of good quality to the customer, but the right first time quality value. In one car plant when this measure was applied the first time quality was measured at around 30 per cent. The aim of this measure was not to embarrass management but to illustrate the hidden loss and provide recognition as it was reduced by improvement efforts. In the process industry where rework and blending of out of spec materials is often built into the process, the quality per cent should reflect this. In a cement plant before this rework measure was introduced it could take hours to produce in spec material because the plant was set up to expect a per cent of recycled materials. By shutting off the recycling flow during start up, the specification was achieved within less than half an hour.

Ideally, quality standards should be set equal to process capability (from assets that are 'fit for purpose'). This level should also exceed what the customer expects. As TQM has taught us (and especially the 'world class' Japanese manufacturers) and what Six Sigma has rediscovered is that perfect quality can be free. Most businesses still believe that there is a trade-off or compromise involved with quality such that to get perfect quality then costs must go up. This is a falsehood. Perfect quality products flowing through processes without interruption will generate more cash more quickly than less capable production systems (Mather, 1988). As will be discussed in Chapter 6, addressing minor product defects is the focus of Quality Maintenance such that, if the production process is in good shape, quality levels will be achieved more easily. Likewise, if defects begin to increase, it is an indication that the process is moving out of control and should prompt timely intervention to restore optimal conditions by engineering, quality and operations staff.

3.7 Summary

Lean TPM provides a framework that is flexible to industry type and organisation structure. It has universal appeal as an improvement process. The Lean TPM approach therefore goes beyond the weaknesses of traditional quality improvements which have tended to lack the full involvement of the company team. The success of Lean TPM, and most other business initiatives, depends on the development of an agreed model of working and a change process to deliver it. To begin this process and model of change management it is important to lower the barriers between inwardly facing business departments (operations and engineering in particular) with outwardly facing departments such as the marketing function. Only by truly uniting these elements of 'world class' manufacturing will the necessary business ingredients and inputs be ready to exploit the opportunities that Lean TPM offers. The model also needs to be a 'living model' that is refined as each layer of waste is removed and incremental adjustments are made to gain stability, perfect the current product line-up and then look into what future manufacturing capabilities are needed to compete in the future. The Lean TPM road map helps to set out the models applied by successful organisations on their journey to world class performance. This leadership provides an important single change agenda across the value generating process. Without such a focus, individual agendas will dilute the improvement process.

Each management level and shopfloor team has a different role as described by the Lean TPM treasure map. The key to success is establishment of a top-down and bottom-up partnership where everyone is committed to finding better working relationships. That is the key ingredient to raising standards and delivering outstanding performance for current and prospective customers to the business and its stakeholders.

In practical terms the top-down role can be defined as:

- Setting priorities (consistency of purpose);
- Setting standards and supporting delivery (collective discipline);
- Giving recognition (objective feedback).

On this last point, studies into teamwork indicate that the biggest failing teams attribute to team leaders is letting people get away with poor performance (Lafasto and Larson, 2002).

The bottom-up role can be defined as:

- Consistent application of best practice (capability);
- Sharing of lessons learned (openness);
- Problem ownership/continuous improvement (aligned goals).

The delivery of these roles is dependent on the joint development of:

- A clear compelling future model;
- A practical change process to deliver the new model;
- Total immersion by management and shopfloor to establish new, more productive working relationships.

Bibliography

Bicheno, J. (2000) *The Lean Toolbox*, 2nd edition. Buckingham: Picsie Books.

Brown, S. (1996) *Strategic Manufacturing for Competitive Advantage*. London: Prentice Hall.

Cohen, L. (1995) *Quality Function Deployment: How to Make it Work for You*. Reading MA: Addison Wesley.

Cox, A. (1996) *Innovations in Procurement Management*. Boston, Lincs: Earlsgate Press.

Christensen, C. (2002) The Innovators Dilemma. *MIT Technology Review* June 2002.

Cummins, C. and Townsend, R. (1999) *Teams in Agricultural Education: An Assessment of Team Process Instruction*. Proceedings of the 26th Annual National Agricultural Education Research Conference.

Kurogane, K. (1993) *Cross Functional Management: Principles and Practical Applications*. Tokyo: APO.

Lafasto, F. and Larson, C. (2002) *When Teams Work Best*. London: Sage Publications.

Mather, H. (1988) *Competitive Manufacturing*. New York: Prentice Hall.

Monden, Y. (1983) *Toyota Production System*. Atlanta: Institute of Industrial Engineers.

Rich, N. (1999) *TPM: The Lean Approach*. Liverpool: Liverpool University Press.

Senge, P. (1993) *The Fifth Discipline*. London: Century Business Press.

Sharma, A. and Moody, P. (2001) *The Perfect Engine: How to Win in the New Demand Economy by Building to Order with Fewer Resources*. New York: Free Press.

Standard, C. and Davis, D. (1999) *Running Today's Factory*. Cincinnati: Hanser Gardner Publications.

Womack, J. and Jones, D. (1996) *Lean Thinking*. New York: Simon and Schuster.

4
Transforming the business model

4.1 Transformation and the business model

As discussed in Chapter 1, many of the 'silver bullets' offered to managers during the 1990s proved to be 'blanks' resulting in failure rather than success and radical business transformation (Hill, 1985). For some reason, many Western firms simply could not sustain these good ideas and proven techniques. But these techniques were not silly ideas – managers are not fooled that easily. These practices were selected to address specific business issues. Their failure suggests that either these techniques were not really understood by the managers that 'bought them' and were not implemented correctly or that the organisation did not have the necessary support structures that allowed these techniques to take root and grow (Womack and Jones, 1996). This chapter sets out the key change team roles to support the Lean TPM improvement process.

In most organisations, working practices have evolved over time, influenced by events which have long gone. A key challenge of the journey to 'world class' is taken with designing a new model and deploying this, without risk, to the business (Hamel and Prahalad, 1994).

4.2 The scope of the change process

To be effective the new business model must include all the activities involving support functions (the 'indirects'), as it is here that many of the business delays and mistakes occur. So it is important to 'vision' what these processes will look like in the future.

For example, a capital project or technical improvement to be delivered in the future will need to be introduced in a way that anticipates and facilitates a new collaborative and team-based approach (involving many departmental stakeholders and the operational teams). This requires the involvement of personnel/training and project engineering as key project leaders. The project might also support changes in the value stream that requires the involvement of the planning department and customer services function (stakeholders). In other words, the impact of some decision taken today will not be felt until later and a simple framework is needed to guide the way. It is important therefore that mistakes, at the early stages, of such a process are minimised to avoid costly confusion, rework and chaos towards the end of the project. The future-state vision is therefore a seamless integration of employees using formalised processes to achieve customer-valued outputs. The ability to set out a 'future-state' requirement for the business and to develop the necessary support activities and education are paramount for 'world beating' businesses. This is not that surprising really as these processes are the key to ensuring that everyone knows where the business is going and how it intends to get there. For traditional businesses, hardship and manufacturing problems are associated with budget cuts and not, as high performance businesses practice, greater and greater levels of training towards the final goal. There are no compromises in delivering the future state, and central to this 'delivery' is getting all levels of personnel to learn, experiment as teams and reflect upon what has been achieved and the extent that the gap between today and the future state has been closed (Dimancescu, 1992).

In modern times, where changes are required on a daily basis, the development of a vision and master plan is critical in determining the direction of change and the velocity of change expected of the firm. Without these two elements, many businesses will meander and fail to make the necessary decisions to transform and improve what they currently manage and treat as 'business as usual'.

Gearing up for implementation involves plotting a course which delivers a single change agenda for decisions taken at the short-, medium- and long-term planning horizons.

In this context, the value of the top-down management role is in planning organisation and control of processes to:

- Establish/raise standards;
- Set priorities;
- Give recognition.

Translation of top-down into bottom-up priorities is provided by a 'change master plan' and the formalisation of the change process and its desired outcomes. Such a 'master plan' is constructed to mirror the stages and milestones of the route map to 'world class' performance discussed in Chapter 2.

These milestones are:

1 Finding the recipe for stable operations.
2 Delivering stable operations and zero breakdowns.
3 Finding the recipe for optimum operations.
4 Delivering and maintaining optimum operations.

Note: By Milestone 2 the business is at a level to gain the PM prize equivalent level of the JIPM and by Milestone 4 will have achieved the level necessary to be considered an advanced PM prize contender.

4.3   Change team roles

Throughout this book we have used the term 'value'. We have used it to describe processes that convert materials into saleable products, to describe processes that compress the time between receiving an order and fulfilling it and defining value from the perspective of the customer. There is however another dimension to 'value' that is rarely discussed in the modern management literature and this is the value of organisational roles within the firm. An understanding of these 'value roles' is an important context that has a relationship with successful and sustainable process improvement. There are nine generic roles which add value to the Lean TPM process.

These are set out in the paragraphs below.

1 *Business Directors and Managers* and their role is to:

- identify winning business and functional strategies;
- support this with a coherent set of priorities and standards;
- consistently manage results and recognition systems;
- to align the organisation and build the capability to deliver them.

This includes key areas of policy such as Focused improvement, Safety, Environment and Administration.

2 *Heads of Operations* whose role spans the planning organisation and control of routine transformation process and delivers Quality, Cost and Delivery (QCD) performance.
3 *Heads of Maintenance* whose role spans the planning organisation and control of asset care and optimisation of current technology. This includes Quality Maintenance as part of the future maintainer role.

4 *Heads of Skill Development/Training Managers* who have responsibilities that cover the provision and operation of systems and processes to raise capability across all levels in the business.

5 *Heads of Technical Management/R&D* whose role covers the provision of technical excellence and flawless delivery of products, equipment and processes through Early Management.

There are two key bottom-up roles within the Lean TPM model and these are:

6 *First Line Managers (FLM)* who lead the shopfloor teams, establishing local policy, developing the potential of the team and its team members and support the continuous improvement process.

7 Multi-skilled *shopfloor teams* whose role is to make good production and to engage in problem-solving where issues are detected with the task or production process.

Finally, there are two key supporting roles including:

8 *Continuous Improvement Manager/Facilitators* to support the planning for, implementation and conditioning of change. Although this role is critical to a successful improvement process, the role has no routine activities. This is a role concerned with building new ways of working, improving collaboration, stabilising the new approach and passing it on. In today's business world of constant change and increasing challenge, the best companies make this a full time role.

9 *Specialists/Key Contacts and Support Functions* that support the value-adding process. Their role is to support the capture and transfer of technical knowledge, target/remove bureaucracy and raise capabilities and create cross-functional collaboration without boundaries.

These are the basic building-block roles of the change team. The enormity of change is too big for an individual to control. In the early stages of developing your own improvement process, such an approach (and the amount of detail involved) can seem confusing and unhelpful. But in reality, there is a logic that underpins a master planning process based on experience of successful change programmes.

It is important to get the basic management tools in place to set out and sustain goal clarity for management and shopfloor worker alike. The following decision steps provide a process to establish top-down roles to support the bottom-up delivery.

Step 1   Define the results required to pass through the next two master plan milestones. This will typically take 3 years or more.

There are many case studies to use as models of realistic, achievable but stretching goals.
For example:

- Deliver zero breakdowns;
- Halve lead times and inventory;
- Minor stops to one-tenth of previous levels;
- Improve quality consistency;
- Reduce new product introduction times;
- Improve morale.

Step 2   Identify the competencies required to deliver that progress. (This should include at least the five classic pillars.)

Step 3   Allocate each competency area to a senior management team member.

Step 4   Define pillars in terms of principles and/or results streams rather than techniques. This is more meaningful to those in the business and it makes it easier to define clear responsibilities and success factors.
For example:

- Continuous improvement in OEE;
- Zero accidents or unplanned emissions;
- Routine operator asset care or self-managed maintenance;
- Effective planned maintenance;
- Continuous skill development;
- Early equipment management;
- Lead time reduction.

Step 5   Identify awareness and training needs including activities to reach consensus on roles and master plan milestone exit criteria.

All of the management team will have an influence on issues of pace, priority and resources allocated and therefore each member of the management team should be engaged with a pillar role.

4.4 Setting and raising standards

A key part of the pillar role is defining guidelines or standards to be developed as local policy by first line management and this tends to include:

- Clarity of goals including the analysis of improvement potential and priorities and provision of information. This includes the use of visual management and two-way information exchange to engage all personnel under a compelling vision and help them to make sense of the world from their perspective.

- Best practice operation definition (e.g. start-up, steady-state and close-down) including cross-shift standardisation.
- Development and application of lifetime maintenance processes to establish basic conditions.
- Skill development and delivery processes.
- Development of a technical trouble map, knowledge base and object-ive decision-making processes to deliver products, equipment and processes so that they achieve flawless operation from day one.

4.5   Implementing 'local' policy

A collective FLM role is the development and enforcement of local 'factory' policy and this will involve some sharing out of geograph-ical improvement zones to create areas of physical responsibility for teams and managers (it may even go as far as colour-coding areas of the factory to denote the team responsible for everything in that area). The development of detailed local policy can then be shared across teams and across shifts to reinforce ownership and responsi-bility of teams and key workers. In addition, the FLM needs to support the evolution of the team towards a desired level of self-management. During this process, the FLM role changes from one of direction to coaching, supporting and finally 'hands off' empowerment. The stepwise development process set out in Table 4.1 guides this task and provides the foundation for self-managed teamwork such that teams can:

- Identify potential improvement opportunities;
- Secure 'zero breakdowns' through effective asset care;
- Optimise process capability through the application of problem pre-vention.

Simultaneously, coaching by senior management supported by facili-tation provides the foundation for FLM leadership development through the stages of:

- Directing the change process;
- Coaching in the new behaviours and responsible empowerment;
- Supporting development of team self-correction capabilities;

Table 4.1 Bottom-up audit coaching level overview

Level	Guiding concept
Teamwork basics A (Milestone 1)	Basic information recording and communication, routine problem definition/frequency and initial cleaning of workplace/equipment, formalisation of critical procedures
Teamwork basics B (Milestone 1)	Standardisation of basic operational and maintenance practices across departments and shifts
High performance Teamwork A (Milestone 2)	Simplify/refine practices to reduce human error/ unplanned intervention and release resource/ energy for improvement activities
High performance Teamwork B (Milestone 2)	Raise awareness of inspection needs to provide early problem detection capabilities. Deliver zero breakdowns and stable operation
Cross-functional teamwork A (Milestone 3)	Build the technical capability of core/workplace personnel and reorganise to delegate routine activities to them. Develop technical team capabilities to focus on optimisation/ stretch targets
Cross-functional teamwork B (Milestone 3)	Bed in new ways of working and identify how to deliver optimum running, improved quality consistency and reduced variability
Cross-functional teamwork A (Milestone 4)	Deliver and maintain optimum conditions
Cross-functional teamwork B (Milestone 4)	Define and strive for next generation of zero targets (e.g. defects, inventory)

- Delegating responsibility;
- Auditing and constructive criticism.

Although often this is supported by formal leadership training and/or one to one coaching, the need for managers to coach their direct reports cannot be side-stepped. Total immersion is necessary to establish new proactive working relationships based on high communication and trust.

In parallel, TPM provides a stepwise bottom-up team-based learning/ development process for shopfloor teams (see Figure 3.2). Combined with the top-down coaching by senior managers and first line management leadership role, this provides the mechanism to deliver progressively increasing levels of team empowerment.

Table 4.1 sets out this stepwise team development and learning framework in terms of the capabilities to be achieved at each stage. In TPM speak, these are the steps of autonomous maintenance. They have been explained in more general terms in Table 4.1 because of their value as a high performance teamwork development process.

4.6 Operations team

The 'role master plan' and bottom-up learning process for shopfloor teams set out what can be achieved and what is desired from each team member within the factory. This reinforces the discipline of:

- Routine operation to make good products and ensure customer satisfaction. This is the primary role of the operations team and its members.
- Improvement/learning from experience.
- Feedback and monitoring so that problems can be escalated by 'managing upwards' and calling for help to eliminate the barriers to 'zero loss' performance.
- Engaging in 'self-management' where trained and safe to do so.

These characteristics are the bare minimum of the operational team member's role. The art of Lean TPM is to increasingly blur the edges of traditional job descriptions such that what is regarded as 'routine' by functional support areas to the operations teams can be packaged and deployed to the team themselves. The movement of quality-assured processes to the team level therefore heightens the sensitivity of the team to abnormalities and increases both skill variety and the empowerment of the teams to act as 'businesses within a business'. Again we return to the softer aspect of managing and the development/integration of new skills via relationship management. We will now explore the major relationships held by the operations teams and support functions.

4.7 Specialists

Team leaders and team members do not exist in a vacuum but instead have a network of support specialists and key contacts upon whom they call when they reach the limits of their knowledge. For most businesses, these *technical specialists* will be internal co-workers in different departments but it is also increasingly true that asset manufacturers and other suppliers are also taking a greater and deeper involvement with shopfloor teams and their training (external specialists). Internal specialist technical staffs hold specific knowledge bases, for which they have attended specialist training/college courses to become qualified in what they do. This is an interesting point to ponder – a maintainer cannot only operate a machine but has received technical knowledge and understanding in how the machine functions etc., and the value of this form of specialist is in using their trained diagnostic skills to help them teach operations staff. Specialists therefore provide assistance in the form of:

- Establishing a 'trouble map' of where efforts need to be introduced to assist material flow performance.
- Horizontal empowerment (standardisation) such that other teams and functions are trained to the correct level of efficiency to do the new roles expected of them.
- To provide knowledge management and training to high standards (including the auditing of standards to ensure they are efficient and safe).
- The 'flawless' implementation of change including designing improvements on behalf of the team to ensure high levels of material flow.

The commercial function and *commercial specialists* also provide the operations teams with a range of benefits and outcomes needed to improve and support the self-management/regulation of team performance. It may be strange to think that this relationship exists as many factory teams have little understanding about the customer but do understand the processing requirements needed. It seems illogical that the 'customer focus' has not been shared to this level. As such, this operations–commercial relationship offers many benefits including:

- Financial awareness so that the teams can make decisions that improve the commercial performance of their area of responsibility.
- Life Cycle Cost Focus to ensure a long-term pay-back to the company that can be used to ensure the 'correct' rather than 'reaction' decisions are made to the production process.
- Cost deployment to focus the improvement efforts of the teams and engineers.

- Customer awareness and 'Voice of the Customer' (VOC) reviews such that changes in the needs of the customer are communicated to the various teams to assist their approach to improvement activities.
- Monitoring the market for innovations that could be employed by the firm including the availability of new technology processes and 'best practices'.
- To ensure that the business is fully compliant with the safety, legal and contractual requirements of manufacturing.

4.8   Facilitation

Having access to technical skills and the willingness of the support functions to assist the development/empowerment process is just one aspect of migrating skills to the person at the point of value-adding. Traditionally, specialists speak with a technical vocabulary and they find it hard (not to say frustrating) when attempting to describe processes and subjects to non-technical people especially operations teams. Without a common understanding then willingness to help will soon evaporate if the technical person is not confident that the operations teams understand and can apply the new knowledge (no matter how willing they are to lose a routine, mundane and non-value-added part of the technician's job). As such, facilitators are needed, and even the smallest of businesses will need a facilitator, to provide the control of the master change plan and ensure that technical and interpersonal training is undertaken to take the teams to the next stage once they have mastered the current stage. Facilitators are therefore instrumental to the management of change velocity and provide the following roles:

- Management of the master plan for each area of the factory.
- Continuous improvement and technical competence needed to train the various factory teams.
- Programme design and co-ordination of change throughout the firm including indirect departments.
- Review the number of suggestions developed by the shopfloor teams and also to monitor the implementation stages of these suggestions.
- Highlight road blocks and support resolution of these issues.
- To maintain the catalogue of improvement activities and maintain all documented procedures (including ensuring these are safe, efficient and are recorded within the quality management system of the firm).

- Maintain a record of training and competence levels achieved by each employee during the change process.
- Provide summary reports for management in terms of progress and improvements achieved.
- Control of training budgets and budgets for external specialist support needed.

Facilitators are also heavily engaged in the promotion of change and the motivation of the teams in the factory as a precursor to sustainable change and, as the name suggests, do not necessarily have to be the experts but must ensure they have a working knowledge of the technique in order to energise the team members and facilitate the change process appropriately. One of the important roles conducted by the facilitator/facilitation team is their responsibility as performance monitors of the change process. As such these individuals are instrumental as a linkage between the management and the overall performance of the new business model in terms of progress. In this manner the facilitation team and factory management are likely to engage in the development of key business reports which will assist the annual rounds of business strategy development and execution. These reports will typically include:

- The development of a 5 year master plan.
- The quantification of business losses and potential.
- The justification and prioritisation of change efforts at the team level.
- An annual analysis of costs including the targeting of costs to be reduced and some influence in the budget process.
- The preparation of quarterly feedback reports which detail the progress of the teams in meeting the business strategy.
- The development of case study learning materials drawn from the teams and used for illustrations for management of the profitability of the change programme (and consequently the benefits of an aligned marketing and operations strategy).

4.9 Summary

The definition of job roles and the assignment of activities to these personnel (to support the Lean TPM change mandate) are critically important in securing a change in the business model. For most businesses some of this infrastructure will exist but has not been 'joined up' in a way that supports 'top-down' strategy deployment and a 'bottom-up' active participation in these changes. A change programme that is not structured effectively will certainly reduce its chances of success in the same manner that a lack of sufficient resources will be a major inhibitor to effective change. Above all, the appointment and training of facilitators is a visible sign that the management of the firm is serious about change and supporting the new business model itself. In short, these features ensure that change is successful and avoids the most basic forms of why programmes fail (lack of time, lack of serious management intent and lack of support).

The analysis of roles and responsibilities is also important in identifying, for each group of personnel, the value they will be expected to play in the future as part of the growth strategy. A perfect strategy will not be realised if the skills throughout the firm are not aligned and focused upon improvements that eventually contribute to the efficiency and effectiveness of the firm, and this demands a rethink of roles. However, as this chapter has portrayed, the roles exist within the firm and don't tend to involve a restructuring exercise (with the associated delay and time lost as individuals adjust and make sense of their new roles as a result of traditional restructuring). Instead, the master plan is a refocusing of what adds value to the primary concern of keeping materials moving, increasing value added and collapsing the payment cycle. These roles exist but, as typical of most traditional management practices, have neither received much attention nor alignment of roles to form a cohesive and robust change management structure.

The alignment of these roles and appropriate skill development, of facilitators and the facilitated, with the future state design of the firm and the development of regular periodic reviews at the close of each Lean TPM milestone ensures that skill provision and the incremental mastery of each Lean TPM pillar maintains an ongoing process of role transition and empowerment of operations personnel (and the empowerment of indirect staff released from these low-value-added routines).

Bibliography

Dimancescu, D. (1992) *The Seamless Enterprise*. New York: Harper Collins.
Hamel, G. and Prahalad, C. (1994) *Competing for the Future*. Boston, MA: Harvard Business School Press.
Hill, T. (1985) *Manufacturing Strategy*. Basingstoke: MacMillan.
Womack, J and Jones, D. (1996) *Lean Thinking*. New York: Simon and Schuster.

5
Process stabilisation

5.1 Stabilising processes

At the heart of Lean TPM is the optimisation of the value stream (Womack and Jones, 1996). As set out in Chapter 2, striving towards the Lean principle of perfection is a systematic journey of learning and developing the organisational capability to:

1 Establish the company-wide best practice recipe for low inventory, high flow stable operation.
2 Lock in the 'best practice recipe' to deliver 'zero breakdowns' and self-managed teamwork.
3 Identify the recipe to release the full potential of the current operation and build the foundation to match and exceed future customer expectations.
4 Change the competitive landscape and lead the customer agenda for products and services.

This chapter concerns the first two milestones of the Lean TPM master plan involving the focus of all employees on 'taming' technology, improving internal value streams, and establishing effective cross-functional teamwork. The achievement of this foundation of robust performance will, as explored in Chapter 6, allow the process of value stream optimisation and innovation to be exploited properly and with commercial effect in terms of generating a means of market advantage (Slack, 1991).

5.2 The recipe for low inventory, high flow and stable operations

Bottom-up and top-down roles

The first side of this organisational 'Rubik cube' is to solve closing the gap between current state asset performance and its reliability envelope or what it could consistently achieve by design. Put another way, the difference between a machine (or process operating) without break-downs and the current state. Getting in control of technology is the first bottom-up target. This provides a hands-on learning programme for shopfloor teams and support functions. It raises understanding of the causes of failure, how to stabilise and extend component life, detect signs of failure early and take prompt timely action. This vastly reduces the amount of unplanned interventions and compresses the time needed to deal with abnormalities as they occur. As each solution is developed, it reinforces learning and the drive for future improve-ment activities because teams can see the benefits of their problem-solving actions.

Activities to raise understanding of value derived by customers, establish current benchmarks for levels of loss/waste and prioritise improvement activities are carried out in parallel. This provides the learning challenge at the management level (the only level that can implement system changes to eliminate organisational and inter-organ-isational wastes). This often overlooked element of the improvement process provides:

- Identification of non-value-adding activities and implementation of low cost or no cost improvements.
- The foundation for the top-down/bottom-up partnership discussed in Chapter 3.

The former activity should improve added value per labour hour by 10 to 15 per cent. This increases the importance of the latter because if these milestones are going to generate 10 to 15 per cent increase in capacity with 10 to 15 per cent less labour intervention, how will management turn this into commercial gain?

Growth strategies produce around five times the return of ones based on down-sizing/head count. In addition, they have a positive impact on motivation. Section 3.1 sets out the leadership challenge and key components. The solution will be different for every organ-isation but without doubt using the Lean TPM process to underpin a customer-driven business strategy is the most powerful way to ensure success.

The route to stable operation and zero breakdowns

The foundation of this activity is characterised by the TPM mantra of *restore before improve*. This principle should be applied to all 'best practice' operations, asset care, workplace organisation and administration activities as well as equipment. From this work, the link between equipment/process condition and overall effectiveness will be proven.

Start with the most critical equipment and processes. Use the process of defining priorities to build ownership to a single change agenda. Include production/maintenance and support groups in the discussion to encourage cross-functional rapport.

Use a simple scheme to assign critical priorities to equipment/ processes such as:

Category A = Critical all of the time/bottleneck;
Category B = Critical some of the time;
Category C = Normally not critical.

When doing this be sure to take into account the future product/business development plans for the next 12 to 24 months. This should present opportunities to integrate the Lean TPM process into the day job. Review existing improvement/capital plans and reassign investment priorities/resources. Separate out improvement priorities into high/low benefit and high/low cost. Give priority to low cost/high benefit activities. In many cases, the insight gained from low cost improvements remove or radically change the scope of technical change required. For example, in a food processing company resources targeted at reducing labour intervention through high cost automation of product assembly were diverted to introducing low cost improvements in mixing processes. Actions included minor refurbishment, the development of best practice standards and operator training. The outcome was more consistent product quality requiring less unplanned intervention during assembly. This released the level of labour targeted by the original project for almost zero capital investment. The benefit in terms of improved team-working and increased confidence to address other hidden losses was priceless.

In short, the goal of this stage of the process is to take action to ensure that:

- Equipment and processes are in a condition where they can do what is asked of them;
- Standards for best practice operation and asset care are in place.

These activities are usually sufficient to deliver 'Best of the Best' OEE performance (see Chapter 3). Typically this is worth a 10 to 15 per cent increase in capacity.

Some managers might question the value of such a simplistic approach and suggest the need to collect data to justify the effort. Experience shows, that for the Lean TPM first step, when the actual costs are compared with the deliverables, the payback is typically measured in terms of months if not weeks. Furthermore, reducing breakdowns has an impact far beyond the immediate cost/benefit results. Put quite simply, there is so much unnecessary waste and cost to remove from most businesses, it is best to avoid the risk that procrastination or 'paralysis by analysis' sets in and nothing happens to change and improve.

The true cost of breakdowns

Imagine yourself as a team leader on the shopfloor when a breakdown occurs: how does it feel to have your plans for the day put on hold and when you have to reallocate resources and spend the rest of the shift/week playing catch up? When such events are common, there is a 'repercussion ripple' throughout the value chain. This 'ripple' is the reason why spare capacity and buffers are designed 'just in case'. Jobs are routinely brought forward to fill a gap in the schedule, materials are borrowed/stolen from other departments, equipment is cannibalised for spares, inventory and customer lead times are increased, overtime is called at short notice. The impact of these daily routine inconveniences is tremendous, such that working life becomes a 'white knuckle ride' for managers and operations staff. It is a ride that soaks up precious management time, stops them from providing value (in terms of planning the change efforts) and loses production capacity that can never be recovered.

Table 5.1 sets out a matrix of action triggers. Those in box 1 are 'firefighting' tasks that must be acted upon immediately to restore a condition of normality. This normality is important because it provides a

Table 5.1 Fire-fighting and time management

		Urgent	
		High	Low
Important	High	1 Fire-fighting	2 Proactive management
	Low	3 Poor prioritisation	4 Political

'yardstick' for operations staff to detect changes and events that throw the production process into chaos and confusion. Think back again to the earlier discussion of demand amplification – breakdowns and constant rescheduling aggravate this situation for your suppliers. When you constantly change orders this information is typically sent to suppliers and, guess what, they see your amplification and react to it. It can be difficult to recover to normality and people become conditioned to thinking chaos is normal, or worse, stop complaining about it and just 'live with it'. This is a poor situation to face – living with it is no longer acceptable so concrete action to restore and then move to the future state is important. If everyone in the factory were to 'blow a whistle' each time things went wrong then ear defenders would be compulsory safety wear at every factory – the trick is to make this whistle blowing the prompt for employees to engage proper and long-term countermeasures (and lower the noise!).

The root cause of management fire fighting is ignoring important Box 2 tasks such as training, asset care, until they become urgent. A significant difference between world class and average companies is the amount of time that is spent in 'box 1' for traditional businesses whereas very little is expended here by the world class. However, for some individuals this is their preferred way of working, a macho environment, and one they have grown accustomed to – indeed, some even enjoy it as it provides an adrenaline boost for them and has been the foundation of their career (the factory 'super hero'). These activities are non-value-adding, they are wastes, and they add precious little to the credibility of the company as a 'world class' manufacturer. They are incorrect behaviours and totally unsuited to the modern empowered workplace and they don't fool customers over the medium term because fire-fighting inevitably has an impact on customer service levels (in terms of quality, delivery or cost performance of the firm).

Constant interruptions to material flow require a series of improvements – there is no single solution or magic cure. Instead, the level of fire-fighting dictates the quantity level of simple solutions needed to recover stability. In this respect it is worth considering: 'What would it be worth to the organisations to achieve zero breakdowns?' and 'How would that feel?' The impact of such a situation has an enormous positive impact upon profitability, business confidence to plan the future and the motivation of the workforce. So stability is not so boring after all and offers a better return on investment, simpler and stress-free jobs and an ability to make dependable promises to customers. These are just the basics of operating a business and it is ironic that many companies aspire to these levels of management control.

To understand why 'zero breakdown' is a realistic and achievable goal it is important to understand a number of key concepts:

- A breakdown is the failure of a component. This is different from a stoppage as a result of a jam, trip or blockage;
- There are only two reasons for breakdowns (poor equipment condition or human error); both of these can be overcome if there is a will to do so through restoration and training initiatives;
- 'Zero breakdowns' does not mean zero maintenance, if the plant condition is sufficient to run without breakdowns for a planned production cycle, the zero breakdown level has been achieved;
- There are world class plants that run for 12 hours or more with zero breakdowns and examples of businesses that have 'believed and achieved'.

Further, significant progress can be made towards 'zero breakdowns' using low cost or no cost refurbishment actions, and much of the restoration/zero loss improvement programme will involve labour hours and time rather than high capital budget expenditure. This time is available for managers who believe the workforce is a fixed cost and not a variable cost (to be fired as soon as possible). The trick is to break the 'old mindsets' which can only be achieved by providing people with the necessary skills and time to improve the operations. Over time, with management actions and support, the old mindsets can be eroded to the point that culture has changed and the deafening noise of 'whistle blowing' will subside. This will be evidenced as the 'overhaul mentality', where parts are replaced just in case, is substituted for data and the working lives of assets extended beyond the point that they have depreciated to zero on the company accounts. Breakdowns are a significant contributory factor to 'box 1' activities – they cause confusion, panic and the adrenaline rush but need to be tamed by the two actions mentioned previously:

- Set and maintain adequate equipment/process condition standards;
- Train personnel/simplify routine tasks to reduce the risk of human error.

Both of these issues are 'box 2' tasks that require leadership in terms of standard setting and implementation by functional and cross-functional management. Establishing standards and making them a habit will not be easy as it breaks the customary practice of operations staff (creating pain). This process is eased significantly when the people who are expected to change have a role in the design of how the systems will be implemented and operated and how best to convince others to think and behave differently.

The gains to the organisation of eliminating these sources of production system 'noise' and chaos include improved consistency, quantity and quality of output, cross-functional teamwork involving learning about how best to self-manage and a reduction in safety risks.

Addressing the causes of sudden sporadic failures, which results in chaotic conditions, is one of the first Lean TPM targets through learning how to set and maintain normal conditions (Figure 5.1). Eliminating high levels of breakdown therefore creates an understanding of normality for both operators and maintainers and it is this platform that increases the value of production exponentially. Such an approach means establishing routine asset care and operations best practice 'as the way we do things around here' and enforcing these rules (taking disciplinary action where necessary for breaches). These routines, through rounds of team-based problem-solving should be reduced to a minimal amount of time needed to conduct the tasks safely and efficiently (moving to convenient points items like machine control dials). In so doing, the facilitators must challenge the teams to find innovative ways of shaving seconds from the process and finding new ways of recording that the tasks have been done (i.e. aircraft style check sheets). Breakdowns are a symptom of poor or inconsistent management standards and priorities, and typically the result of short-term cost down focus (Figure 5.2). Weaning the organisation away from this outlook takes time but can be achieved if a consistent approach and standard practices are replicated across the entire factory.

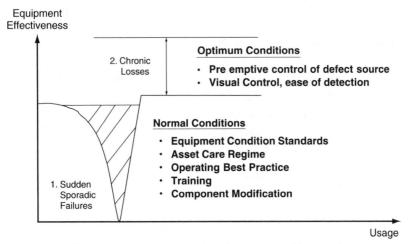

Figure 5.1 The route map to zero breakdowns and beyond

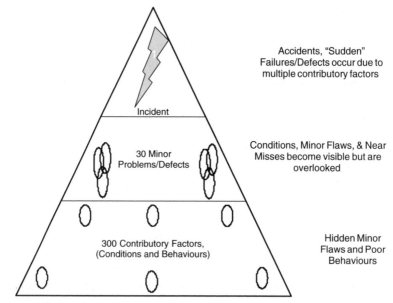

Figure 5.2 Patterns of accidents, breakdowns and defects

5.3   Setting work standards

The nine-step process of TPM sets out a proven route map to help management to kick the habit of fire-fighting and establish 'normal operating conditions' (see Table 5.2). *Note*: Developed by WCS International as a framework for applying a multi-discipline teamwork approach to TPM. Within this framework, measurement activities (steps 1–3) are carried out in parallel to steps to improve equipment condition (steps 4–7) so that the improvement can be measured but also information is provided for targeting future improvements (steps 8 and 9). The measurement activities include the mapping of the value stream and are dealt with later in the chapter.

The outputs from the nine-step process are practical standards which if applied consistently will help shopfloor teams to deliver the Lean TPM Zero ABCD goals:

- Accidents;
- Breakdowns;
- Contamination;
- Defects.

Table 5.2 Setting work standards

Step	Activity	Reason
1	Collect equipment history and performance information	Identify key sources of information and make them available. Collate those technical, commercial and operations records which are available
2	Define OEE measurement and potential	Establish key performance measurement, improvement potential in terms of its impact on the value stream
3	Assess the hidden losses and set improvement priorities	Identify base case levels of hidden losses, establish priorities and improvement potential
4	Criticality assessment	First understand the equipment, how its functionality impacts on key criteria such as safety, quality and reliability and set standards for equipment condition
5	Conditional appraisal	Clean the equipment and inspect every square centimetre using agreed standards. Look for sources of accelerated deterioration and scattering of dust and dirt
6	Refurbishment	Restore the equipment condition and minimise the causes of accelerated wear where possible
7	Asset care	Formalise best practices activities for correct routine operation incorporating operator asset care and planned maintenance
8	Best practice	Simplify asset care and operating best practice into a working routine
9	Problem prevention	Target remaining areas of loss to achieve normal conditions

The route to delivering each of these is raising work standards and making these a habit for operations staff. Only then will managers and team leaders be able to establish behaviours where abnormalities can be detected and dealt with to maintain asset and material flow control.

Techniques such as the nine step process helps managers to set policy standards and teams develop these into detailed work standards. For example:

- *Policy standard* (set by management): Welding equipment will have a routine of daily cleaning and inspection checks;
- *Local standard* (set by shopfloor teams):
  - change CO_2 wire;
  - check pressure setting;
  - clean table and tooling;
  - check clamp head security;
  - check for air and water leaks;
  - check torch and harness security;
  - etc.

Ideally, multi-discipline shopfloor core teams will include technical specialists to ensure the quality of work conducted. This process will set the standards rather than process engineers on their own (the old way of imposing standards without team 'buy in'). Involvement of this nature has a number of benefits including:

- Increasing shopfloor understanding and with it ownership;
- Builds closer working relationships between functions;
- Ensures that the standard reflects the workplace reality;
- Ensures that training material is written from the perspective of those who will complete the task;
- Once a team have been shown how to set a standard, they can produce others and refine the standard to reflect changes in product/process specifications;
- The process engineers are no longer the rate limiting factor to the standard setting activity.

Naturally, standards set by shopfloor teams would need to be subject to a suitable technical change process and a review of safety procedures before these new standards are authorised and recorded within the quality management system (document reference numbers and review procedures established). The best medium for achieving the seamless introduction of such standards is through the use of Single Point Lessons. An A4 sheet of paper (with digital pictures showing good and bad practice) for operators to use like a motorcycle 'Haynes manual' to control a process or task. These documents are invaluable sources of knowledge and standardisation which 'world class' businesses can learn well and practice extensively.

5.4 Leading the implementation of standards

Once standards are set, implementing them and making them a habit takes leadership. Traditionally, in hierarchical organisations, shopfloor workers have been conditioned to receiving change rather than taking ownership for improvements. Supervisors have been conditioned to become co-ordinators of the shiftly/daily 'white knuckle' ride: making sure that workers are occupied, materials are available for the next run, arranging overtime. Improvement in this environment is driven by initiatives, often involving 10 per cent or less of the workforce.

To stay out of 'box 1' (Table 5.1) for ever means achieving 100 per cent involvement and harnessing the innovation of the multi-discipline teams. In Lean TPM this is achieved through a team development process (see Figure 5.3) designed to meet the challenge of achieving

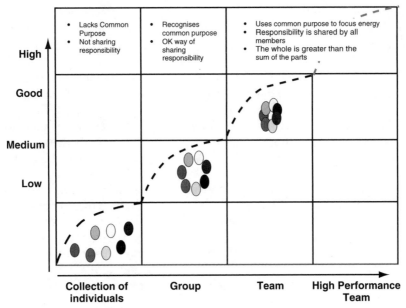

Figure 5.3 Improving team performance

stable operations and zero breakdowns rather than through external outward bound training.

One of the most powerful tools to introduce and reinforce such behaviours is known as CANDO or simply workplace organisation. Originally known as 5S after the five Japanese words used to describe each step of the workplace organisation process. The letters CANDO spell out the first letters of English terms for the same process (Cleaning, Arrangement, Neatness, Discipline, Order). Although these are not faithful translations of the Japanese words they are a useful acronym of the process steps. This provides a simple process for teams to take ownership of their workplace and learn how to set and implement common work standards. Table 5.3 sets out the CANDO steps in terms of goals and improvement themes for each step. In the table the corresponding Japanese term, which may be used on other literature/case studies, is given for those interested in understanding the origins of this thinking.

The CANDO steps can be applied to all manner of workplaces. It can even be used as a set of rules and routines for keeping your computer

Table 5.3 Workplace organisation (5S and CANDO)

Step	Goal	Improvement theme
1 Cleaning (Seiri)	Get rid of what you don't need	Nothing in the workplace which we don't use every month
2 Arrangement (Seiton)	Arrange what is left so that it is where you need it	Find everything you need in 30 seconds or less
3 Neatness (Seisi)	Make it easy to keep the workplace clean	Fix the routine for keeping the workplace neat and tidy with minimum effort
4 Discipline (Seiketsu)	Maintaining a spotless workplace	Understand how to reduce contamination/non-conformance and make it easy to do right (keep to the standard)
5 Order (Shitsuke)	Apply visual controls	Make it difficult to do wrong

workspace (disk) clear of unnecessary items. Working with teams to introduce CANDO reinforces a number of positive behaviours:

- How to establish rules/standards which the team will abide by;
- Empowering the team to take control of their environment;
- Setting standards which make it easier to produce good work.

On study tours to exemplar sites, the most frequently asked question is what you would do differently if you could start again? The most frequent answer to that question is that they would spend more time on CANDO/5S workplace organisation activities. The CANDO process is therefore an important activity. The state of the factory workplace discipline is also a good indicator (and measure) of factory morale. When morale is low there is stress to maintain discipline and when it is high the factory should sparkle as a showroom to customers who may visit. Keeping the factory in 'battle ready' condition should also be subject to rounds of problem-solving, simplification and reducing the time needed for such cleaning activity. For many businesses, especially those without a formal process of problem-solving, the CANDO activity is a great way for employees to learn about problem-solving and innovation before being taught about it and applying it to machinery and processes. In effect, the CANDO activity uses a latent ability to solve problems concerning an issue that is close to the hearts of all employees. It should be noted that the close relationship between CANDO activity and workplace safety makes it a change initiative where early trade union involvement (in management and auditing of performance) is desirable. Safety is foremost in trade union thinking and developing the change mandate to include a role for the trade union is well worth having as it reinforces the legitimacy of the programme and assists with its policing.

5.5 Establishing operator asset care

Table 5.4 sets out how the CANDO steps are applied to equipment/processes. This equates to the first four steps of autonomous maintenance. *Note*: The Autonomous maintenance is one of the core TPM principles developed by JIPM. Although this concept is commonly shown with seven steps, the first four support the goal of zero breakdowns. The remaining three support the achievement of optimum conditions and are dealt with in Chapter 6.

Achieving and maintaining 'zero breakdowns' is possible but not easy; it requires constant vigilance. With business routinely introducing

Table 5.4 CANDO for equipment and processes

Step	Activity	Improvement theme
Cleaning	Clean the equipment and restore functionality	Cleaning is inspection. To see what is happening
Arrange routine tasks	Formalise procedures and countermeasures to common problems	Detect problems and understand the equipment principles and improvement process
Neatness	Set cleaning and lubrication standards	Fix the routine for keeping the equipment neat and tidy and in good working order with minimum effort
Discipline	General inspection	Understand how to reduce contamination/non-conformance and make it easy to do right (keep to the standard)
Order	Apply visual controls	Make it difficult to do wrong

new products and equipment, requiring greater flexibility and higher levels it is easy for standards to slip. Like the golf pro who practices their swing daily, the maintenance of normal conditions takes discipline and a true commitment to high standards. CANDO provides the foundation for that discipline and commitment. What's more, if standards start to slip, you can see it. If visible standards start to slip, imagine what is happening to the discipline for processes that are less easy to observe.

Achieving normal conditions provides a predictable pattern of output. Although zero breakdowns is a 2–3 year goal, the benefits can be felt at bottleneck processes in less than 6 months. What you get is increased capacity and quality consistency. If the additional output can be sold, unit costs will improve. If the additional capacity cannot be sold, on its own, the benefits are significantly reduced. How should such potential be harnessed so that it is translated into value that the customer is prepared to pay for.

5.6   Understanding the voice of the customer

'World class' organisations are outwardly focused and understand the customers they serve and hope to serve (Womack and Jones, 1996). The best way to conduct this analysis is to bring together all the functions and departments that must combine to deliver a typical order from the point of customer interaction to the final despatch of the product (a door-to-door approach). Such an approach requires the combination of representatives from the sales functions, the operations, quality assurance, maintenance, production planning and supplier scheduling at the very minimum. Each of these managers has, to a greater or lesser extent, an influence upon the performance of the firm and the service delivered to its customers. It is also these business functions that need to be brought together to act as a single decision-making team that designs the future capability of the firm. It is also this team, through a process of mutual dependency, that must understand both the external needs from the market as well as the internal demands and constraints facing all other departmental managers. No modern world class company has ever gained this title with a fragmented organisation that defends insularity and departmental boundaries.

World class organisations have also learned that the 'command and control' structures associated with the mass production era are no longer sufficient to increase the decision-making capability and responsiveness of the firm. As such, world class organisations design systems and do not change any part of the production system without understanding the impacts of these changes on all other parts of the firm (Figure 5.4).

With this team in place, finding out the major customers to the business is relatively straightforward. Gaining an insight into the most important customers can be gained from a simple pareto analysis of annual

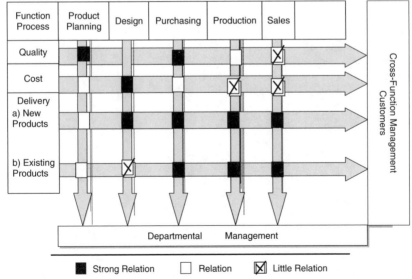

Figure 5.4 Cross-functional team process impact

volumes bought by all the customers of the firm (Figure 5.5). To understand these main process flows and their associated products is important in order to gain control of the products that represent the 'runners' (high volume) and raise the most income for the business. There is little point in beginning a process of stabilisation by starting with low volume products (this can come later). Initially the focus of Lean TPM will be the main products, customers and support processes needed to produce the mainstream stock-keeping units. These products also recover, due to their volume, the most of factory overheads.

Having identified the customers to the firm, and before any internal analyses are conducted, it is important to determine the buying criteria of these businesses. For many firms this is the first time that managers, from many internal departments, have gathered together to understand the true 'voice of the customer'. Instead, in industry it is often the case that only the marketing and sales functions collect customer information and tend not to share it with those managers charged with converting materials into products that customers want and are willing to pay for. The problem with both the TPM and lean approaches is that all too often, and despite an implicit acknowledgement that satisfied customers result from the actions of many business departments, the majority of manufacturing businesses still don't share this type of information.

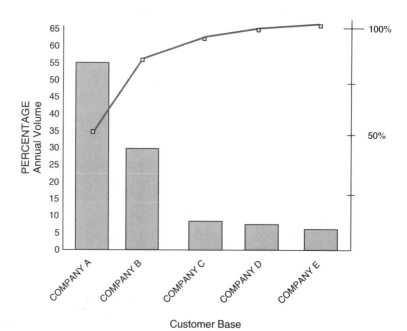

Figure 5.5 Pareto analysis

As such it is no wonder that process stabilisation rarely occurs and the first 'rule' of lean thinking is treated as a marketing not a business priority.

Finding the 'voice of the customer' and translating it into a value proposition demands questioning and seeking out the things customers want and clearly prioritising these needs in order to build a production system that is capable of consistently achieving these needs. For all customers to manufacturing firms these needs can be summarised into a range of features, many of which go well beyond the mere notion of 'price'. To begin the VOC process it is worth thinking about two aspects of 'value', identified by Hill (1985), as:

1 What output performance features are needed to 'qualify' our business to engage in trade with the customer base? These features are the basic implicit and performance features needed simply to meet current customer expectations of good service.
2 What output performance features will be order winners by differentiating our offering from its competitors? These features include providing customer order qualification features at much better levels than the rest of the competition and industry as well as innovations or extras in the product-service bundle offered to existing and new customers.

For the purpose of process stabilisation both order qualifiers (the most basic of expressed customer needs) and order winners represent a solid foundation upon which to build a high performance value stream that gives customers their needs at a profit to the firm. Most customers value basic outputs in the form of a stated level of quality performance, delivery performance and price. These can all be plotted and as all manufacturers know, it is possible to create production systems that can reconcile these demands. For instance, to build a production system that offers the highest level of quality will reduce costs (due to less failures, better due date performance will result and productivity will increase). However, it is also the case that new 'buying' criteria such as environmental performance, may also be stated (often an order winner for customers keen to show their concern for the environment). At the heart of these debates will be quality, delivery and cost (QDC) as the basis of customer service. In modern times, it is delivery performance that has yet to be fully mastered and optimised to allow good quality material flow to be combined with an ability to reduce costs and lead times. This round of preliminary analysis is very useful and will generate a series of performance indicators that allow the firm to benchmark with its chosen sector (as well as identify a number of existing and new customers who would be attracted by performing at this level) (Figure 5.6).

Customer Want	Score/10	Rating Against Competitors Bad OK Good
Improve quality of product/service	8/10	
Make dependable delivery promises	5/10	
Offer durable and reliable products	10/10	
Offer fast deliveries	6/10	
Make rapid volume changes to meet demand	2/10	
Lower lead times	10/10	
Offer customisation	7/10	
Offer rapid design changes	9/10	
Offer full product availability	10/10	
Introduce new products quickly etc.	2/10	

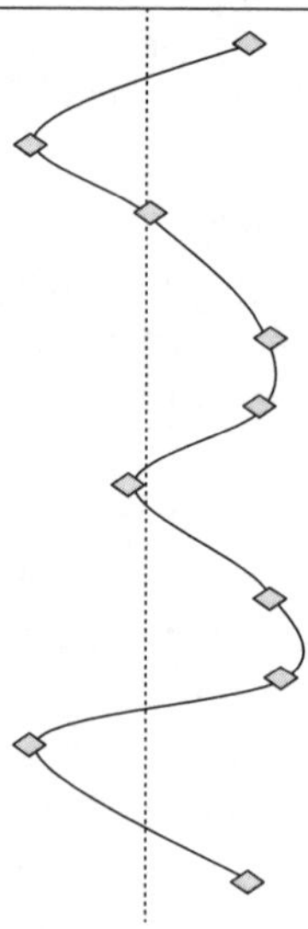

Figure 5.6 Competitive analysis

However, understanding the current 'voice of the customer' is not enough and, given the length of time needed to change a production system, it is necessary to debate and discuss the trends and future predicted performance requirements needed by the existing customer base. This round of discussions merely identifies what performance is needed to continue to qualify for work with existing customers and therefore should also include discussions of what the optimal level of customer service would be. A note of caution must be sounded here. It is not desirable to think of these order qualification and winning criteria over a greater period than the next 3 to 5 years because any greater length of time is subject to vagary and many imponderables (including things like regulations, new technology, etc.). Thinking over too long a time frame is unlikely to yield meaningful information with which to reconfigure and optimise the internal processes of the firm.

At this stage it is important not to get bogged down in debating the minutiae of future predicted customer needs. Consider broad improvement themes to support the development of in-house capability such as:

1 Zero defect quality products leaving the factory with zero losses to quality through the process. Whilst many manufacturers will have achieved a zero defect rating with their customers, few have achieved

it at the internal process level and instead rely upon inspection, by officials or operators, to 'filter out' process defects. Even companies at 'parts per million' (PPM) levels of performance often find that the internal production process has not mastered quality performance to the PPM level.

2 A 100 per cent due date performance in a lead time that is half of today's level or stated customer expectation as a minimum.

3 A reduction of production costs and prices charged in the region of 3 per cent per annum.

4 A greater range of products and a much shorter time to market (from the drawing board to available to order).

More detailed VO Customer profiling is applied during the optimisation milestone as discussed in Chapter 6.

5.7 Visualising the value stream

The next stage of Lean TPM is to visualise the value stream through which value must be generated without waste, delay or interruption. This again is an exercise well worth conducting with the entire representative of middle managers whose functions influence the quality, delivery and cost performance of the firm.

The objective of this stage is to draw the basic blue print of the manufacturing process (and its associated information flows) as a series of connected diagrams (Figures 5.7–5.9). Having conducted these analyses and 'maps' the next stage is to identify the problems with the existing system and create a series of potential future state manufacturing models. The combination of these maps provides a benchmark of what currently happens and what the optimised manufacturing process value stream could be like. The latter is obviously the focus of the continuous improvement activities of the firm and is based upon the performance measures that have been predicted in terms of basic qualification criteria and the order winning level needed to maintain a 'manufacturing-led competitive advantage' over other producers in the same sector. The process of mapping the value stream involves:

1 Identifying the main manufactured products by the factory in terms of a pareto rating, and importantly the main products bought by the major customers to the company that were discussed in the previous stage. This analysis provides an understanding of A Grade products

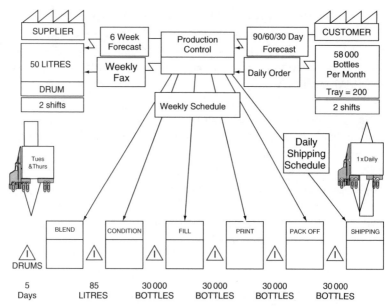

Figure 5.7 Current state map

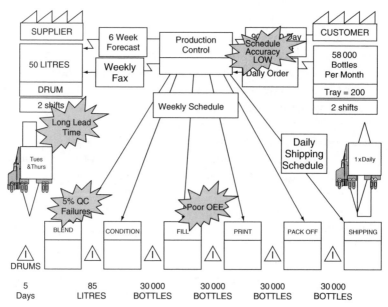

Figure 5.8 Problem map

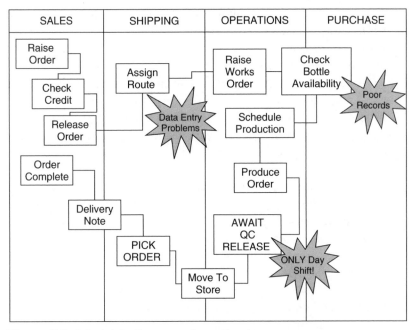

Figure 5.9 Administrative mapping using brown paper

(those products with the highest annual sales and regularity of production), B Grade products (medium sales and production regularity) and C Grade products (the tail of the product offering with very low volumes and infrequent demand). A general 'rule of thumb' is that the A Grade products will account for 50 per cent of all products but 80 per cent of annual volume whereas the C Grade products will account for a lot of the products but only a very small percentage of annual sales). This analysis is critical (even though it is typical for all these types of product to be subject to the same quality, delivery and cost expectations). The purpose is therefore to design a production system that allows the 'A grade' products to be manufactured with zero losses and let the other products benefit from this process design.

2 To list the sequence of manufacturing stages through which the product is made and converted into a finished good and to identify, for every stage:

- The available minutes in a working day;
- The cycle time of the asset;
- The minimum and typical batch sizes processed by the asset;
- The changeover time required in changing between products.

3 Draw, at the bottom of a landscape page of paper, these stages and their associated data boxes.
4 Identify the typical customer orders, schedules and delivery patterns in the top right hand section of the map.
5 Identify the main supplier (key item of the bill of materials) and the typical amount of purchasing and delivery patterns in the top left.
6 In the middle at the top, plot the order processing procedures and join the work schedule to the work station sequence of operations that have been written across the bottom of the chart.

Having written up the basic 'current state' performance map it is now possible to overlay problems and constraints of the 'as is' production system. You may find the use of a number of overhead projector acetates useful as a means of overlaying the problems on to the current state map. These problems should be analysed, for every process stage and information channel, in terms of:

1 The process with the longest 'total cycle time' (the combination of the cycle time multiplied by the batch size and the time needed to changeover the asset) should be clearly identified. The process with the lowest average daily production output potential should also be identified. In most cases this will be the same process area but it can vary due to differences in shift patterns.
2 Problems affecting the quality of production at the stage in terms of physical losses.
3 Problems affecting the delivery of materials such as the availability of items such as packaging or the lengths of the shifts in each area.
4 The cost per hour of operating the manufacturing stage showing the costs of using the workstation to produce outputs.
5 Problems affecting the flexibility of the process stage to manufacture a variety of products including the identification of processes with long set up times.
6 Now look for opportunities to improve the system through simplification, a combination of activities that can feed each other directly and the elimination of activities that add no value (to the business or the customer).

In parallel, the same analysis should be conducted for the top of the map and the information exchanges to identify what aspects of the order cycle can be eliminated or compressed to result in less lead time needed to get customer orders into the physical process of manufacturing. A recent study conducted by the Lean Enterprise Research Centre

(Cardiff Business School) found this analysis to represent a significant source of loss and variation to automobile manufacturers (representing many weeks of activity but with just a day of time involved in the assembly of a vehicle (Holweg, 2001). To conduct this analysis it may be necessary to decompose the information flow to assess how orders are processed and through which departments.

All too often, these hidden manufacturing processes represent procedures that have remained unchallenged even during the rise in popularity of quality circles. As such many computerised planning systems contain obsolete data and processes. For instance, many manufacturing businesses update their computer planning system weekly and as such have added a week to the lead times of all suppliers. On other occasions you may find that traditional safety stocks, entered when the system was being introduced, have rarely been questioned or that supplier lead times have changed but these have not been incorporated into the computer system data bank. As the *Harvard Business Review* article stated, managers must pin themselves to a customer order to understand what happens within their own system and the appalling number of hand-offs and delays experienced even before a production order is printed and given to the operations departments. You will be unpleasantly surprised to find out how long an order will wait before it is converted into an instruction to manufacture product.

5.8 The process of stabilisation: The free-flowing materials map

The third stage is to ensure work flows, just like a liquid through a pipeline, across the internal value stream and then through supplier companies in order to service customers. To understand the 'flow performance' of the current state production system it is necessary to apply the OEE measure to each process stage and record the result on the map itself. The most important of these OEE calculations is therefore at the bottleneck operation which dictates the actual output levels of the entire firm and it is this OEE which will be the focus of all improvement activity. It should again be noted that to focus all activities upon the improvement of this OEE measure has commercial benefits but to improve this figure may necessitate improvements at other points in the value stream. The logic is quite plain to see:

1 The bottleneck operation determines the output performance (what the customer sees in terms of the sequence and availability of outputs).

2 It is necessary to ensure that all work stations after the bottleneck do not generate quality losses (as these are very expensive and have used bottleneck capacity which is the most limited resource).

3 The sequence of materials getting to the bottleneck is important because this determines the availability of work at the bottleneck. Thus if the bottleneck output sequence determines the time at which goods are finished it is important that pre-bottleneck processes have good quality and internal due date performance and manufacture in a batch size that closely matches that of the bottleneck.

If a 'total production system' approach is taken to manufacturing and logistics improvement then all processes should be targeted with improvements – if only to ensure that they do not become a bottleneck. For example, an investment decision to buy a second asset in the bottleneck area may push the new bottleneck process to another part of the factory at which it will be difficult to improve unless work has already begun to achieve a zero loss status at all non-bottlenecks. Think about it – it makes commercial sense to improve performance everywhere as any sub-optimal performance does in fact offer no true value to customers but does add costs. It is important to remember though that the bottleneck process does have a major impact upon the commercial performance of the firm and also the flow of materials within it. As such, bottleneck management will dominate the design of the production system during the debate on how to get production flowing.

The second phase of analyses is to assess the eighteen losses as set out in Chapter 3. Here the management team must analyse the entire supply chain process and the wide variety of wastes that exist within it. Finally, to add more insult to the injured egos of managers it is worth calculating a few additional measures of performance. These include:

1 Multiplying together the average quality and delivery performance of the firm (both expressed as percentages) to its customers to the day stated by the customer. If quality performance was 98 per cent and delivery on time was 70 per cent then this would yield 98 per cent times 70 per cent or a performance figure of 68.6 per cent. Now the average number of finished goods held (in hours) should be calculated.

2 The internal OEE of the production areas.

3 The supplier performance should be measured in the same manner as the firm's performance to the customer and in the same time

demands (i.e. to the day of requirement). Here performance is again determined as the average quality performance multiplied by the average delivery performance (both expressed as percentages). Thus if quality performance is 92 per cent with 60 per cent on time performance then supplier performance to yield is 55.2 per cent.

This form of diagnostic analysis really cements an understanding of a production system which each manager works within, makes decisions but often sub-optimising the production system by optimising each business function and forgetting about the value stream. It is the stability and improvement of the value chain that provides the 'customer qualification' outputs needed by customers. It is also possible that this form of analysis yields the conclusion that free flow of materials in the current state is not possible and that the production system is under-buffered. The latter is usually the result of many rounds of pure internal cost cutting rather than system design and management using the key processes of quality, delivery and cost performance. Now it would be naïve of us to argue or propose that the production system can simply be changed in a short time frame and without problem. This would not be a good reflection of what the management of the factory has been doing until this point of attempting to optimise the value chain! So, practically there may be some solutions that still need to be found. In this case it is important to still continue with the improvement planning but revise certain aspects of the production system design.

To ensure a good level of flow, and where it is physically possible, any production stage that is followed by another with a shorter overall cycle time should be combined. Similarly, where flow is not possible, due to a number of reasons that will have surfaced during the mapping of the value stream, it is necessary to determine the points in the production system at which an amount of controlled inventory is needed to buffer the system. These buffers serve to protect the production system from disruption, the impact of high variety, machinery that constitutes a system bottleneck or other constraint. The buffers (investments of company cash in terms of stocks) are used to allow work to be pulled by demand and for manufacturing businesses to collapse their lead times to customers. In reality, these buffers are quite financially cheap compared with the performance of the overall value stream, and have the impact of slowing stock turns from an accounting point of view but commercially strengthen and support flow performance throughout the value stream. In short, costs will not be minimised but the quality and delivery performance of the production system will

benefit. If introduced the buffers allow the flow of products (as measured in terms of OEE performance) to be stabilised and this approach
allows the factory management to revisit the need for buffers at a later
stage. If you prefer, this approach is similar to using safety stocks to
cover against problems whilst they are sorted out and then, at a later
stage, the safety stocks are removed without risking the performance
of the system.

At this point the real improvements to the production system can be
understood and the process for optimisation can take shape. These
debates will tend to include the feasibility (given the amount of stock-
keeping units of the company) of using:

1 A total push system whereby orders are loaded and progressed
  through the manufacturing facility via the use of a schedule.
2 A total pull system involving the use of deliberate buffers of standard product to allow the finished goods of the firm to be lowered
  until a set amount is reached and internal re-orders and replenish-
  ments of the 'like' product is made. This fills back up the finished
  goods available for sale. As such, finished goods pull like products
  from the finishing operations and so on to the initial processes of
  the production system.
3 A hybrid approach is used where pull is introduced for production
  centres of families with low product variety or buffer spaces are
  used and withdrawals trigger the release of the next job in the manufacturing schedule. The latter is often associated with the management of the bottleneck operation and the subservience of all other
  manufacturing process stages to the schedule of this process. It
  should be noted that the bottleneck determines the overall capacity
  of the production system. The hybrid system therefore seeks to protect
  the operation with a buffer of either standard or scheduled products.
  An hour lost here can never be recovered and is lost to the whole
  production system whereas an hour lost at a non-bottleneck can be
  recovered as the non-bottleneck has the ability to produce more
  products quicker than the bottleneck.

Finally, once the basic production system has been created and stability restored within the factory, improvement activities are needed to
maintain high performance as well as ever increasing the capability of
the asset to result in flow without buffers or waste. The latter is
reached when the bottleneck OEE has risen and stabilised at a consistently high level of performance and the target for improvement is the
'door-to-door' OEE measure.

These analyses will, just as during the 'voice of the customer' debate, unearth a whole host of issues that impact upon flow of materials to meet customer needs. These debates aimed at reconciling the design of the production system and the target performances needed to 'qualify' and 'win' orders over the next 3 to 5 years will include some suggestions for improvements. Starting with the internal value stream (within your factory) and the current state map produced in the previous section, it is possible to identify and cost the major projects that will release improvements in efficiency to increase the effectiveness of the firm and its ability to meet customer qualification criteria. The best way to achieve this form of systematic change and improvement is to combine an approach to cost deployment with an X-chart approach to justifying the expenditure of investments on production system improvements. These two powerful tools, used to quantify and focus the improvement effort are invaluable when it is necessary to justify the use of limited business funds and to allow all managers to 'buy into' why the money is necessary. Such an approach is not typical of existing Western manufacturers who tend not to think about production as a system and often attempt to improve little bits of the production system that have absolutely no impact on the flow of materials through the bottleneck process.

5.9 Locking in the recipe for low inventory, high flow operation delivering zero breakdowns and self-managed teamwork

So far in this chapter we have set out the tasks of establishing the recipe for zero breakdowns as a bottom-up process and mapping/ stabilising the value stream as a top-down process. In reality, both activities are integrated. The value stream mapping activity will highlight critical and bottleneck processes. The improvement in reliability will enable the removal of buffers and revision of planning standards.

The exit criteria for the first of these two stabilisation milestones as set out in Chapter 2 are:

- All personnel are actively involved in the improvement programme;
- First line management ownership is beyond question;
- The gaps to deliver stable operation have been identified and a clear sequence of actions is under way to deliver that.

The title of this section sets out the goal for the next milestone. The same techniques are applied as in the first milestone but as knowledge improves and routines are formalised and visual management techniques are applied they can be simplified.

Milestone 2 Refine best practice

This presents the opportunity to transfer roles which had previously been considered specialist without fear of quality problems. Doing so requires training in underpinning skills but as breakdowns are reduced it becomes easier to make time for such activities. Known as Horizontal Empowerment this process progressively releases specialist resource to focus on the future milestones of optimisation. In doing so it provides the opportunity for up-skilling of all functions including management (Table 5.5).

Under this process the operations team can be developed to have the capability to carry out all routine activities. This means that they can be made accountable for delivering the 'zero breakdowns' goal because it is within their capability to deliver it. They can then be given recognition when it is achieved. It will only be achieved if:

- Management provide the structure, systems and organisation;
- Team leaders provide discipline, consistency and manage results not tasks (feed forward);
- Teams have the capability, openness and goal clarity.

Exit criteria for this milestone are:

- Use of feed forward mechanisms rather than feedback;
- No recurring problems;
- Stable delivery lead times;
- Routine tasks are carried out by self-managed teams/cells.

Table 5.5 Horizontal empowerment

Current role	Future role
Operator	Technician
Maintainer	Engineer
Supervisor	Team manager
Manager	Entrepreneur

5.10 Summary

Process stabilisation is the first significant milestone on the route to Value Stream Perfection (Womack and Jones, 1996). Achieving a 'restored normality' clears the technology landscape of the debris of poorly designed systems and working practices that lead to inevitable fire fighting and customer service failure which act to ransom customer service. These first steps are as much about dealing with gaps in management processes and priorities as they are about innovation and smart technology. It is about leadership in terms of consistent direction/priorities, raising standards and releasing potential, and it is about formally writing down processes and procedures so that everyone can understand the system. In truth, most manufacturing systems are the amalgam of lots of peoples activities and specialist knowledge but it is only when you chart what happens on a day-to-day basis that you understand the rate (or lack of) flow. Putting all these issues on a single map helps highlight the issues – much of which is 'low hanging fruit' and needs only a time investment. Just imagine what an organisation can achieve when all this fire fighting is sorted and is the exception rather than the 'norm'. For companies where every day is a 'white knuckle ride' this may seem far-fetched and unachievable but in 'world class' companies it should be noted that this is not the ultimate goal but just the entry ticket. The use of a current state and future state map is priceless in this process of visioning the future factory and showing employees the damage caused by processes that are out of control. Further still, it will awaken a commercial thinking that traditionally has not been part of workforce involvement. Imagine the horror as a machine operator, who experiences hassle on a daily basis to maintain performance from the machine, discovering that the company has 3 months of finished goods stock of a major item. This awakening is ideal and inevitably creates a sense amongst employees that 'if this was my business then I'd run it differently'.

When you have achieved this level of understanding then the battle for 'world class' performance is underway and people's hearts and minds are aligned in a correct manner. This scenario is far removed from the traditional factory systems where an operator arrives at work for another mundane session of 8 hours moving bits without thought. Now you face a workforce who will question and innovate. Most managers will find this, at first a bit daunting, even threatening to their management practices, but this will 'wear off' as this intent is constructive critical rather than personal. After all, managers work within the system and try to do the best they can just like every other employee.

The future state map should therefore be designed to allow managers to manage the big business issues and work with other managers whilst the shopfloor teams take charge of what they are best at – producing and improving the factory's material flow systems.

Having gained a position of 'normality' by stabilising the production system the improvement effort will slow unless managers and facilitators take the next milestone steps to improve the operations of the production system. The next level, to perfect the current set of technologies and working practices, is covered in the next chapter.

Bibliography

Hill, T. (1985) *Manufacturing Strategy*. Basingstoke: MacMillan.

Holweg, M. (2001) PhD Thesis, Cardiff University.

Slack, N. (1991) *Manufacturing Advantage*. London: Mercury Press.

Womack, J. and Jones D. (1996) *Lean Thinking*. New York: Simon and Schuster.

6
Process optimisation

6.1 Introduction to the challenge

In Chapter 5 we set out how the dual targets of 'zero breakdowns' and 'stable internal value streams' provide the entry ticket to World Class Performance.

At this point, managers who sit back and congratulate themselves are similar to the hare in the children's story of the race with the tortoise. The hare loses due to its complacency and arrogance. It got half way round the race but did not finish. In the same way management lose interest in completing the marathon of Lean TPM implementation but there is still much to be achieved and enjoyed. Some managers justify this decision because they feel that the smaller potential gains are not worth it (almost like a decay curve has set in). Unfortunately though, the key to consistently high customer service is completion of routine tasks without error. These tasks become boring if the drive for improvement is not present. Small problems appear to be insurmountable, standards slip progressively resulting in death by a thousand small cuts. Alternatively, continuous improvement can reinforce the disciple of routine operations so that as each layer of waste is removed, not only are costs lowered but new opportunities become available to:

- Increase capacity at relatively low cost;
- Improve quality consistency/reduce variability and the risk of human error;
- Get closer to the customer;
- Raise skills and with it the motivation of the workforce.

Avoiding the management decay curve is important and the challenges of these new milestones are great and offer the most reward – especially for managers.

Together these new challenges enhance the potential of competitive advantage as long as it is supported by vision and leadership. No one ever said that creating and sustaining a lean business was easy and if it was then there would be no competitive advantage to be gained by doing so. It is no wonder then that only around 1 per cent of companies achieves true 'world class' levels of performance. It is only these few that have achieved the ability to set the trend of change in their chosen market and to do so in a manner that commands customer loyalty.

Behind the difficulty of improving business performance for competitive advantage, through a manufacturing-led advantage, lie a number of major issues. Professor Yamashina, an internationally accepted TPM guru, outlines these by proposing:

1 In the West managers don't tend to train operational staff in the functionality or technical knowledge concerning correct operations of productive assets.
2 In this situation, mature continuous improvement activities have eliminated the low hanging fruit of poor materials supply, untidiness and lack of standardised work. Improvement teams may then identify problems/opportunities to improve assets but without basic engineering skills, they will not be able to resolve them.
3 The inability to solve engineering-related performance problems necessitates a greater integration with the engineering departments (training/support) which results in frustration for operational staff if these issues are left unresolved or poorly supported.

According to Yamashina, the result is a glass ceiling for skills and improvement activities. This is a serious issue and goes directly to the heart of developing a self-sustaining company-wide Lean TPM improvement programme. Application, application and application is the answer to sustaining change but what is important at this stage is defining the right questions to ask of the production process and ideal working conditions as much as asking questions about what the future employee should be capable of achieving (Figure 6.1).

There are a number of reasons why this glass ceiling can be so difficult to breakthrough. Few of them are technical, most relate to human behaviour. The paragraphs below set out the key factors to be overcome.

6.2 Changing drivers

Organisations which have become used to fire-fighting, often relax their efforts as they approach zero breakdowns. The feeling that the fire has been put out will lull the unwary into a very false sense of

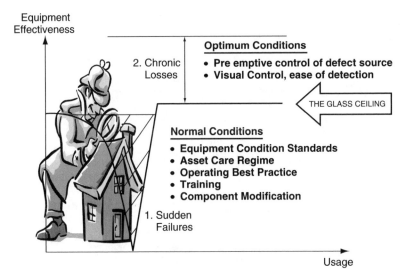

Figure 6.1 The route beyond zero breakdowns

security. When management fire-fighting is the norm, there is a certain satisfaction in getting through the day unscathed. Take this away and the working environment can seem very dull. It is not unusual for organisations that achieve stable operation to feel that order volumes have dropped when they are actually processing more business with less effort. To hold the gains, such businesses need to learn how to be driven by what is important rather than what is urgent. The implications for the organisation can be far reaching. As shown in Figure 6.2, the application of Maintenance is one of those important tasks which can become urgent if not addressed. The application of Maintenance can be likened to that of the gas supply in a hot air balloon. On the way up, the flame can be reduced and the balloon will keep rising. Once the balloon begins to fall, if the gas to the flame is not increased, it will reach a point after which the balloon will crash no matter how much gas is applied. This point of 'no return' is not obvious but it is none the less certain. Learning to operate in the important/non-urgent sector requires a different sort of outlook and judgement. If the organisation has not learned this capability, it will not progress past the glass ceiling. This can be quite a difficult skill for FLM to acquire after years of learning how to make do and mend.

To stay with the analogy, an organisation, which is skilled at repairing crashed hot air balloons, will tend to develop an organisation, systems, capability and culture which supports this. To progress further, the organisation needs to develop new capabilities and a new management

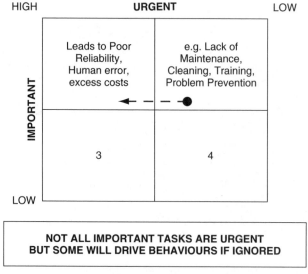

Figure 6.2 Behaviours reinforcing reactive management

model. Milestones 1 and 2 provide the foundation to progress from fire fighting to a stimulus based on engagement with a compelling vision. In Chapter 6 we focus on the process of transformation from one successful business model to an as yet undefined future model. Making progress as identified in Chapter 1, requires both leadership and management to deal with the issues which reinforce limiting behaviours (see Figure 6.3).

Internal systems and procedures, organisational structure and reward/recognition are the domain of management. The language and folklore, accepted patterns of behaviours and relationships/power structures is the domain of leadership.

The leadership challenge is to create the compelling vision behind the business case for raising standards, pressing further with the Lean TPM principle of customer focus and engaging the passion to improve further. This includes proactively creating the culture you need, not leaving it to chance.

The Management challenge is to establish the hard-edged discipline to remove fixing and fiddling in favour of structured improvement based upon data and targeting. Truly 'world class' managers have this intolerance to a level where they cannot hold back from exposing poor practices and standards which are not good enough – not for the traditional dispensation of blame but simply that these events are learning points for everyone. If your future state organisation has the belief that

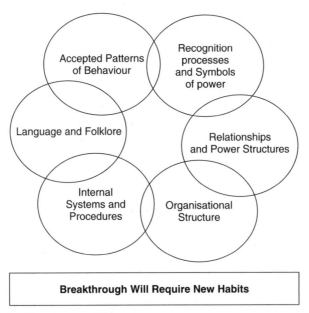

Figure 6.3 The glass ceiling

employee integration is a must, then you have no option but to subscribe to the 'no blame' culture. Blame at this stage will be regarded by operations teams and others as a return to the 'old ways'. Further, these events and intolerance are not a sign of arrogance but simply the observations of 'trained eyes' that can see the value in removing waste in a process even though it is basically stable. Put simply, the next two milestones shift the focus from problem-solving to problem or opportunity finding in pursuit of customer/market leading capability.

What gets in the way?

In Chapter 2 we set out the route map followed by organisations that have successfully broken through this 'glass ceiling'. In each case, these organisations have been driven by the passion to develop and implement strategies which *require* higher levels of effectiveness and which have the attraction of delivering outstanding competitive advantage. This has been a key part of their breakthrough process. Not all of them made it at the first attempt. Figure 6.4 illustrates the dilemma of those initial failures.

These companies recognised that the passion for improvement feeds on challenge and that setting world class challenges for factory teams

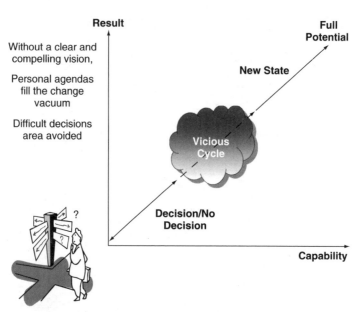

Figure 6.4 Breaking through the glass ceiling

helps to accelerate the empowerment process (providing they are also given the means and ability to deliver the challenge).

Take, for example, the vision of a process plant that had already made major improvements in zero breakdowns – the challenge was to double again the time between stoppages so that they could run with confidence through the night (and avoid having the burden of high overtime bills or the third shift for parts of a year). In addition, the completion of this challenge provided the 'business winning' ability to take customer orders up to end of the working day for next day delivery. This could hardly, given what we have said before, be seen as a boring challenge to keep people going – it was instead a significant addition to the competitive arsenal of the firm.

A similar approach was used by the manufacturers of seasonal products so that they could add or reduce the number of shifts as demand fluctuated, without adding or laying off personnel. The biggest benefit was the ability to service peak demand direct from production without the predictable reduction in quality caused when hiring temporary labour. Again, the significance of this 'pursuit of perfection' can never be overstated. The result is a fitter and leaner business that is capable of opening new markets as much as it can now defend itself during 'price' competition (and without giving away valuable margin!).

What distinguishes the managers in these companies from all others is their ability to think and act strategically and improve the *production system*, rather than being driven by short-term mechanistic cost-reduction exercises and interfering in a piecemeal manner. Such short-term thinking puts head-count reduction high on the list of cost-reduction targets which is illogical and counter to the beliefs of 'world class' companies. In world class businesses there will always be the need for people in transformation processes so they keep job descriptions 'loose' to allow redeployment of displaced workers within the factory processes and they rarely lay-off because to do so can ransom the improvement process. The latter is an important issue, and if workers belief that their improvement ideas will harm the livelihood of co-workers then innovation will stop dead – 'no one wants to improve a mate out of a job'.

Care and attention is therefore lavished upon the operator grade so that they can play an increasingly important role in the management, control and improvement of technology. Reductions in unplanned intervention means operators are not the 'output rate' limiting step and multi-skills ensure people can move fluidly to where they are needed in the process to improve or stand in for others. As such, to move to an optimal state of production performance, the operator is trained to take on new roles including:

- Routine maintenance and administration;
- Dealing with peers within customers and supplier organisations supporting:

  o The introduction of new products/services and materials;
  o Developing innovations to reduce total supply chain cost.

- Simple repairs and technical skills in the workshop;
- The instruction of operators in terms of how to visualise asset deterioration (heat-sensitive strips etc.).

These are but a fraction of the opportunities for new role development available for businesses that have gained stability. Many choose to engage operators in additional qualifications, such as the National Vocational Qualifications (NVQs), and most will seek out a greater relationship with a local technical college in order to devise 'company-specific' courses delivered to employees at the place of work. As the operator role matures, the role of the maintainer changes, away from fiddling and fixing breakdowns towards an engagement in optimising the process. The challenge to optimise is a welcome one for the maintainer, if a little uncomfortable, in that it calls for the application of the 'technical skills' learned at college. For the team leader there are also benefits especially as 'time' becomes available to manage the medium-term business issues and engage in business development activities through cross-functional groups of other leaders. If this sounds exciting it is. If it sounds far-fetched, you should be aware that the two companies (mentioned above) had achieved this state and perform at extremely high levels of OEE (and D2D) since 1997. Unfortunately, both are in Japan but could so easily have been located in the UK if businesses had engaged in Lean TPM sooner or some of the TPM pioneers in the UK had not given up or had their specialist facilitators poached by others.

6.3 Springing the strategy trap

In summary, moving past this potential watershed depends on the development of a business vision and strategy which can inspire the whole business; one which can progressively replace the fire fighting action triggers with ones which secure progress towards a collectively worthwhile goal. Many alternative futures are available to those who achieve milestone 2 stable operations. Time taken to make a sound choice is time well spent but it is important to recognise that analysis will only take you so far. After all the options have been evaluated, the brutal truth is that when embarking on a new strategy, all of the details cannot be known. If this were not so, there would be no point in trying to compete.

Basing strategic thinking on past/current business models will reinforce the status quo. When the market shifts and the business model no longer delivers results, few clues will be offered about what to do next. The business graveyard is littered with examples of business models which once worked but then failed. Springing the strategy trap requires a different outlook. Figure 6.5 sets out this important change of focus: one where predictable external pressures are anticipated. For example, interest rate and currency movements are predictable, what is not known is the timing. As mentioned in Chapter 3, research shows that 10 to 15 per cent of customers will change supplier due to circumstances outside of their control. A further 45 per cent actively review their supply options each year. It is predictable, therefore that some customers will be lost. Robust strategies are flexible to such changes because the lead team have considered their response to such events and can move quickly when they occur. This strategic assessment of business scenarios should not result in analysis paralysis but

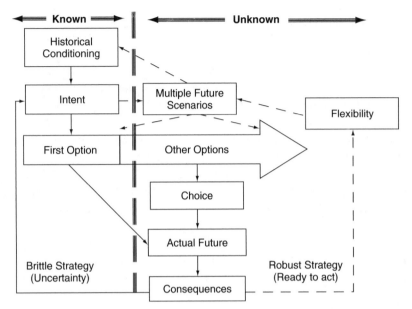

Figure 6.5 Springing the strategy trap

should raise awareness and interest in the suitability of available strategic options so that the organisation can keep these open as long as possible.

The importance of this deeper level of thinking ('the one business' approach) cannot be underestimated because without it, there remains the risk of a 'knee jerk' response to a shifting market by cutting head count (and slicing out employees with value-generating potential). As well as the loss of future opportunities, redundancy scenarios leave a lasting scar on those who work in the organisation and again reinforce the view that management cannot be trusted or that they are incompetent and cannot be relied upon.

Organisations will not break through the 'glass ceiling' without trust, mutual respect (that does not have to include liking management!) and a feeling that there is a co-destiny between the survival/growth of the firm and the individual employee. This isn't to say that in certain circumstances head count reduction is not appropriate but it is exercised fairly and undertaken to ensure the survival of the business or to compensate previous poor decisions which have left the organisation vulnerable. Lean TPM is a tool to help management avoid such mistakes by focusing business activities on growth and maximising all resources to keep costs at a minimum.

6.4   Creating flexible organisations

A common weakness of traditionally structured hierarchical organisations is the way in which individual business functions operate in isolation and indeed set their own strategies rather than being an integrated
and aligned element of the business. Such poorly aligned businesses
can be successful even though they retain an insular departmental
focus (even keeping this view during milestones 1 and 2). There
comes a point when the lack of business process thinking and 'business level focus' creates problems with decision-making speed, when
these self-centred structures limit the pace of change and improvement
(Hammer, 1996). Reaching the point where these departments must be
integrated is easy to detect, it's the point at which progress towards
improvement goals stagnates and political infighting fills the vacuum.
Here businesses will lose the good and mobile people unless a new
focus is provided; one where customer value is maximised through
collaborative internal projects to improve business performance.

Moving beyond the 'glass ceiling' depends on the evolution of an
organisation which supports such collaboration to secure horizontal
empowerment. As mentioned in Chapter 5, this occurs when routine
processes are simplified and what was previously a specialist activity
can become part of the organisation's core competencies. For example,
once equipment is in good condition and the activities needed to
maintain this condition are standardised, operators can carry out routine maintenance because they are closest to the process, and with
training, can identify when intervention is needed. Eventually the
operational team will develop the capability to deal with all routine
activities areas such as planning, quality control and internal logistics. This is an important mechanism for releasing specialist capability to support delivery of the new strategy. The low level of
interaction in hierarchical organisations inhibits this process. Figure 6.6
provides a model manufacturing organisation defined by three broad
functions.

Removing boundaries between them can only be achieved where
there is a deliberate policy to perpetually up-skill workers of all types.
After all 'teaching a man to fish so that he can feed himself' is the
approach that must be taken to skill development. This means that
organisational structures need to be flexible enough to cope with blur-
ring the edges of these overlapping functions to encourage higher
levels of integration and support rapid change. However, this capability to
involve indirect functions is not at the expense of removing these functions altogether. They provide the resources and capabilities to reach

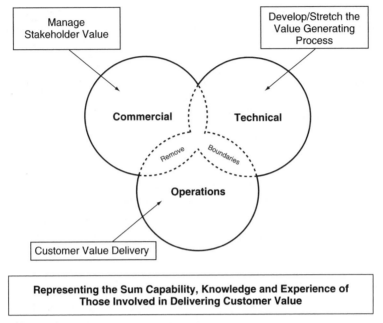

Figure 6.6 Horizontal empowerment

new heights of business performance. This is the core enabler for the delivery of outstanding results from within the organisation.

6.5   The optimisation process

Milestone 3: Build Capability (Extend Process Flow)

The aims of this milestone are to identify the recipe to release the full potential of the current operation and build the foundation to match and exceed future customer expectations.

This can be said to have been achieved when the organisation is fully engaged in the task of delivering customer leading capability, characterised by the following exit criteria:

- Have identified critical optimisation targets (commercial, operations and technical) and making progress towards them;
- Maintained zero breakdowns whilst transferring routine condition monitoring and servicing from maintenance to production personnel;
- Established a clear product/service 'innovation stream' strategy capable of achieving customer leading performance;

- Focus shifted from internal improvement to include external partners;
- Technical focus shifts to external scanning for asset innovations rather than internal correction. Engineering staff to ensure that they are included in any future asset specifications at the capital expenditure and procurement stages. Feedback by improvement teams to this knowledge base is routine. This includes intelligence about improvements needed for the next generation of assets.

Top-down process

Given that the programme scope includes the Leadership agenda, the generic management focus for approaching the next dimension of the Lean TPM 'perfection' Rubik cube is 'Optimisation' to deliver multiple 'zero targets' and not just breakdowns but all forms of loss. Here, loss modelling and cost deployment are useful tools. Loss modelling is used to identify the links between cost and hidden losses. It provides a top level prioritisation process which is capable of considering complete supply chain optimisation. This is important as, for example, activities to improve flexibility, shorten lead times and reduce WIP/Finished Goods inventory levels can result in increases in raw material stocks. Here sub-optimisation of one part of the chain helps to deliver lower overall costs. Another example, the outsourcing of cleaning tasks may make sense in isolation but can increase maintenance costs and production downtime through conflicting priorities. Without the precursor of loss modelling, cost reduction can constrain supply chain optimisation.

As part of this review consider each of the nine elements of the business value generation map (Figure 3.4) taking into account current capabilities and future needs. Using the hidden loss treasure map (Figure 2.8 and Section 3.2), assess the impact on business performance of the future state value stream map at 100 per cent supply chain effectiveness. If this seems too fanciful, set what you believe to be a practical 3 to 5 year improvement goal (for example 150% of current effectiveness levels). This assessment should quantify the benefits of increased sales volumes as well as reduced direct costs per sales unit.

Assess each area of the treasure map in terms of potential contribution using a simple 1 to 3 scale where 1 is no impact, 2 is some impact and 3 is significant impact. Use this assessment to apportion the estimate of overall benefit to the most significant areas of hidden loss. Allocate a champion to review each loss area in detail and identify current costs and potential benefits. Figure 6.7 shows a typical output from such an exercise.

Cost Deployment Summary

Location: Engineering Company
Version: 1.0 11 April 02
Top sheet attached ☐Y☐N
Scope of system: Component Production
Potential Gain: £560,500

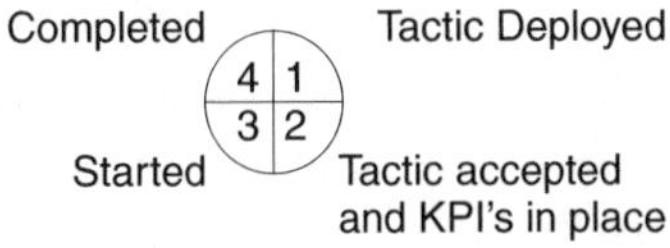

Completed   Tactic Deployed

4 | 1
3 | 2

Started   Tactic accepted
and KPI's in place

No.	Tactic Description	Cost	Forecast Benefit	Resp'	Status
1	Improve OEE to release capacity for new business development. (Overhead reduction)	£10k	£181 500	AF	⊕
2	Standard shift working (Contributes to 7) to raise lowest shift productivity to average	£5k	£68 000	PM	⊕
3	Standardise Planned maintenance and carry out refurbishment to reduce sporadic losses by 25%	£20k	£100 500	AB	⊕
4	Refine/training in core competencies to improve flexibility and reduce avoidable waiting time (Contributes to 2)	£10k	£150 000	CW	⊕
5	Improve Bottleneck resource Scheduling to reduce avoidable waiting time by 50%	£5k	£54 000	MN	⊕
6	Reduce human Intervention during equipment cycle to improve productivity (Contributes to 1)	£15k	£20 000	RL	⊕
7	Improve best practice and technology to halve the quality failures	£5k	£40 000	RL/PM	⊕

Figure 6.7 Cost deployment summary example

In this way, the loss modelling helps to identify priorities for improvement and provides a key part of the cost deployment brief for an appropriate multi-discipline team(s). It also provides the mechanism for setting goals and monitoring progress (typically quarterly). The team's first activity is to carry out a detailed cost audit to confirm the level of saving achievable and a programme to deliver it. Experience shows that such teams normally increase the level of benefits identified by the initial loss model.

Once the costs deployment process is in place this becomes a two-way process with teams identifying their next improvement target as part of the delivery of their current goal. When integrated with the business planning process this provides the delivery mechanism aligning accountabilities behind a single change agenda.

Bottom-up focus

At this stage, typical aims are to reduce the need for intervention during production runs, to achieve vertical take off of quality following change-overs, to stabilise and extend component/tool life to deliver flexibility and outstanding performance. The main targets for the optimisation processes are:

- Zero contamination (and the need to spend excessive time cleaning);
- Zero defects (and the requirement for technical functions to prevent the manufacture of defects during manufacturing).

Although those of you who have progressed down the Six Sigma route may be concerned about the 'concept of zero', here we use it as a targeting exercise which leads to the progressive creation of optimal operating conditions. In this evolution some zero goals are easier to achieve than are others.

In a plant making steel pipes the achievement of zero pipe jams resulted from fitting a mirror so that the furnace could be continuously observed without the discomfort of looking directly into the oven. Within a very short time the conditions which created 'jam ups' were identified and the problem was consigned to the history books. The benefits to the plant and personnel were enormous.

The most important first step was engaging the workforce in the 'zero goal'. The process engineers had already failed to solve the problem with a more technical approach, the idea for the mirror came from a shopfloor worker. The movement towards thinking about optimal conditions is a watershed for the Lean TPM organisation. It marks a transition from event-driven 'Five Why' thinking, the science of determining optimal conditions. This 'Five Hows' approach uses proactive versus reactive thinking which demands good technical knowledge and a positive learning environment.

Zero contamination

Zero contamination is an important step to achieving optimised operations. Dirt and dust getting into moving parts will vary component wear rates and make component life difficult to predict. Much of the easy to deal with contamination sources will have been dealt with through the actions described in Chapter 5. Common actions include:

- Implementing equipment modifications/design standards to minimise scattering of dust and dirt;
- Establishing maintenance standards for reassembly of ducting/dust control mechanisms following maintenance;
- Working methods which avoid the use of airlines to blow out contamination.

Making significant progress towards zero defects will generally mean raising standards and introducing a new generation of countermeasures at the machine and with the operations teams. In some industries, the benefits of doing so have more than justified the cost of introducing 'clean room' workplaces to replace previously uncontrolled working environments. Eliminating the need to clean therefore recovers time and productive capacity which can be usefully sold in the form of increased outputs and this is an important element of combining a growth strategy with cost minimisation.

Zero defects

The main driver for the optimisation process is defect reduction. Figure 6.8 sets out the main generic causes and examples of the parameters to be optimised.

The outcome of 'zero breakdowns' does not mean 'zero maintainers' but instead provides the opportunity to transform and redefine their role within the business to increase added value. Typically, this also includes training of operators and maintainers in quality control tools and mistake-proofing devices. Value added in the sense of the maintainer role is easy to understand – it is the collective diagnostic skills of these workers – after all that is why these individuals went to college and took all those exams. Within the TPM Quality Maintenance pillar, maintenance personnel have the tools to release value from current processes by focusing on the occurrence of minor defects. They also provide the bridge to improvements in the specification of future equipment supporting the Early Management Process discussed in more detail in Section 6.6.

Optimise	Typical Parameters
Variation in Operating Environment	Temperature level, Temperature change Humidity. Shock, Vibration Incoming Material Variation, Voltage change, Batch to Batch Variation, Operator Intervention
Potential For Human Error	Visual controls, Standards, Activity Indicators, Alignment markings, warnings, Andon, Audible controls, Mistake proofing
Equipment/Process Deterioration	Corrosion, Brittleness, Dirt and Dust in moving parts, Perishing, lubrication, Alignment, Oxidisation Piece to Piece Variation, Process to Process variation, tooling storage and Retrieval damage, Instrumentation calibration, cleaning processes

Figure 6.8 Zero defects and optimisation

Using the five hows

The process stabilisation activities described in Chapter 5 focuses on problem-solving and dealing with sporadic losses. As previously mentioned, this process is characterised by the five whys technique to find out *why* something happened. Optimisation is concerned with problem prevention and chronic losses.

Chronic losses have multiple causes that are not always transparent at the time. In fact, chronic losses can often become sporadic problems given time. Dealing with them is characterized by an approach to identify *how* the impact of contributory factors can be reduced or removed. Table 6.1 summarises the steps of the five hows technique.

Table 6.2 provides a checklist of points to consider when defining optimum conditions.

A useful tactic to expose latent conditions of an asset is to change the process conditions to which it operates which could include increasing speed of the process to magnify the contributory factors to quality defects. Experience shows that in some cases, the chronic loss cause/effect mechanism can be unclear despite significant analysis. In such cases it is not possible to clearly identify what is triggering off a defect. Identifying the likely contributory factors, defining and implementing standards to reduce their impact provides a practical progress towards the zero defect goal. In some cases it is easier to begin from the perspective of addressing sub-optimum conditions first before beginning the analysis. For example, Figure 6.9 describes a common problem on a bottling line where caps fed from a vibrating feeder jam causing intermittent supply.

For the technical and operations teams, it was unclear what had caused each jam and there seemed to be no common pattern of failures. Previous

Table 6.1 The 5 hows approach

How	Notes
Define 1. *How* and where the target defect occurs	Identify the characteristic of the defect by defining it in physical terms, e.g. dimensional precision, outer appearance, assembly precision.
Identify 2. *How* the defect can be brought under control	Identify the processes that could contribute to the defect. Understand the function of each and assess the conditions that will minimise causes of variation.
Design 3. *How* to control and 4 *How* to reduce defect levels	Observe/experiment to understand which parameters have the most impact on target defect levels (80 per cent of the control is with 20 per cent of the potential parameters). This often needs to include parameters which have not previously been considered as important e.g. heat or humidity, method variation. Introduce low cost or no cost ideas first. Aim to first stabilise and then reduce the defect levels. Focus on ease of detection and early response.
Refine 5. *How* the controls are applied to make it easy to do right, difficult to do wrong	Refine the process to reduce the need for technical judgement without reducing levels of awareness and understanding.

attempts at problem-solving had resulted in ad hoc adjustment to feeder vibration settings and a return to the traditional 'fiddling' of machinery to try and reach an acceptable outcome. Following the 'Five hows' methodology, the process of brainstorming commenced with asking 'how' five times to reach zero loss performance. Here is what the team found, just one illustration of the many they originated and investigated, and how they ascertained the optimal conditions for equipment operation:

Zero Jams of Bottles during Processing

1 *How* to define the problem in physical terms (friction of bottle on process conveyor).
2 *How* to control the level of friction. In some instances it is necessary to define a precise measure on the phenomena to be achieved.

Table 6.2 Considering optimal conditions

Factor	Potential improvement targets
Operational conditions	Processing conditions, operability, conformance to specification
Assembly precision	Precision of assembled parts, vibration, assembly fixtures, interfaces/linkages/timing
External appearance	Dirt, scratches, rust, pinholes, deformation, discolouration, seizure, uneven wear, cracks
Installation precision	Vibration, level/fit
Function	Operation across complete range, compatibility with other parts, actuating systems, intrinsic reliability
Dimensional precision	Required precision, surface finish, life span
Environment	Dust, dirt, heat, cleaning process, pipe work, waste disposal routes
Materials/ Strength	Specification tolerance, storage condition, ageing/ brittleness, corrosion

In this case it was sufficient to look for how friction might be reduced.

3 and 4 *How* to control and reduce friction (countermeasures that could be introduced). These countermeasures included:

- Alignment of the vibrating feeder and slide to the filler;
- Routine clean out of the vibrating bowl feeder to reduce the build up of plastic debris on the slide;
- Cleaning of the slide once per shift;
- Setting a maximum level for loading of the vibrating bowl feeder.

The outcome was that, through the combined team effort of applying operator knowledge and technical skills, the countermeasures reduced the jams to zero. The team later found another source of problems (plastic debris eroded from the bottom of the sacks which the caps were delivered in that contaminated the conveyor lines) and adopted the same approach to optimising conditions. Neither of these solutions was expensive and was insignificant in comparison to the benefits delivered. Finally, the teams 'closed out' the problem-solving by documenting the approach as a case study.

BIRDS EYE VIEW of PROCESS

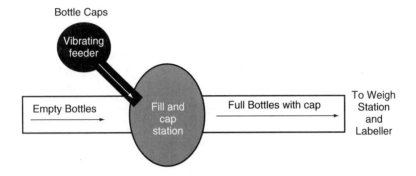

Chronic Loss: intermittent feed of caps to the filler

Frequency: 10 times per shift. Mean time between intervention
4.8 minutes

Impact: when caps are not fed, line stops and needs to be reset.

Figure 6.9 A bottling line case

5 *How* to make it easy to do right and difficult to do wrong.

- Visual indicators on the hopper levels and next to the cap storage area. Single point lessons adding the clean out of the bowl weekly and the slide (daily), painted match lines used to highlight bowl/ feeder chute alignment.

Having learned the link between defect prevention and optimum conditions, the team was able to transfer that lesson to other 'problem areas' with great effect and almost zero cost.

Tools to support optimisation

Although the five hows require a different outlook in terms of problem definition, the nine step process and value mapping tools set out in Chapter 5 are still applicable to the process of understanding and resolving optimisation issues. The major difference in approach, during the optimisation of equipment, lies in the selection of improvement priorities. These are therefore driven by a clear understanding of how the customer makes decisions and what the customer values.

The ultimate goal of this stage of the journey is to deliver market/customer leading capability. To support this a more detailed VOC analysis is used to provide the context for product and service development. This includes information which will provide an insight into many more issues that customers take into account when buying products. The categorisation of features is as follows:

- *Qualifiers Features* which the customer does not discuss but expects to be offered (these are the most likely causes of customer complaints). Such features include the implicit understanding that the product is safe and will be of good quality.
- *Winner Features* which guide customers' buying decisions and these are features they value and are prepared to pay (more) for. These features are used to compare between competitor manufacturers and are a means of differentiating your product in terms of better performance than the others.
- *Excitement Features* that reinforce customer loyalty by providing hidden value. Such features could include extended warranty for products or the ability to recycle the packaging used.

The analysis also provides guidance on unmet customer needs. To quote Michael Dell 'One of the magic abilities of any great product company is to understand technologies and customer requirements and come up with the perfect combination to solve a problem. Customers sometimes do not realise that they have a problem to solve. The customer is not likely to come and say they need a new metallic compound used in the construction of their notebook computer but they may tell us they need a computer that is really light and rugged. Where some companies fall down is that they get enamoured with the idea of inventing things and sometimes what they invent is not what people need.'

Voice of the Customer profiling also identifies the sources of competitive drift/advantage that could be exploited. It also provides a measure of current customer relationships and how to improve them and a structured template for comparison of the current/planned product and service portfolio. Finally, it helps to predict and protect the 'value' offered by the business to its customers.

To illustrate the VOC categories. A tablecloth in a high class restaurant is an example of a qualifier feature. It is expected and would only be noticed if it was not there. Satellite navigation in cars is an excitement feature and is typically something that would not be the reason for buying the car but would be a welcome feature. Excitement features have a habit of becoming winner or qualifier features over time. At one stage a remote control TV was an excitement feature, now it is very much a qualifier feature. Winner features are those which the customer uses to distinguish between options. In theory these are the features which the customer would be prepared to pay more for: higher top speed, wider screen, haut cuisine. It is interesting and profitable to use this form of analysis with teams and to identify the control items on the equipment that must be carefully managed to ensure that qualifiers continue to be met as customer expectations increase. The same goes for winners.

Finding out how customers weight each feature (prioritisation by the customer) is achieved by sets of paired questions. For example if a customer is asked:

- A: How would you feel if the feature is present?
- B: How would you feel if the feature is absent?

If his answer to A is not bothered and his answer to B is unhappy, this is a qualifier feature like the tablecloth in the restaurant. If his answer to A is good and B is not bothered, this is an excitement feature like the satellite navigation. A typical VOC exercise would be carried out with each main customer category to identify:

- Where low scores on qualifier features indicates a risk to competitiveness;
- Where the potential winner features indicate the opportunity for additional value; and
- Where 'excitement features' suggest potential unmet needs to use as the focus for innovation generation processes/new product development.

These features can then be analysed to identify:

- How to define the feature in operational/technical terms;
- The link between the feature and the process used to generate it;
- The relevant process variable/parts (control points);
- Control points for those process/part characteristics.

The X matrix (see Figure 6.10) provides a means of representing and understanding these relationships. This example shows an assessment of the interaction of processes which impact the delivery of the key risk/opportunity features from a VOC study. This provides the first step towards understanding what options are available to optimise process parameters. The matrix also illustrates how the investigation process has been split across multiple functions to facilitate innovation and opportunities for improvement. Although a lead function is assigned to each control area, implementation of improvements will be

Figure 6.10 The X matrix

Feature × Process (upper section) — grand total 40

Feature		Mix 28%	Shaper 15%	Dry 20%	Kiln 23%	Pack 15%
Cosmetic Appearance	Basic	2	2	3	3	3
Predictable Life	Perf	3	1	2	3	1
Long life	Perf	3	1	2	3	1
Strength	Perf	3	2	1	3	1
Total		11	6	8	9	6

Centre X quadrant labels: **Feature** / **Process** / **Control Point** / **Parameter/Part**

Parameter/Part × Process (lower-right section)

Parameter/Part	Mix 28%	Shaper 15%	Dry 20%	Kiln 23%	Pack 15%	Total	
Raw Material Quality	2	1	1	1		5	10%
Dust Control	2	1		2	2	7	14%
Cutting		3	1	1	2	7	14%
Start up routine	3	1	2	3	2	11	22%
Steady state routine	3	2	5	3	3	16	32%
Close down routine		2	1		1	4	8%
Total	10	10	10	10	10	50	

Action × Parameter/Part (lower-left section)

Parameter/Part	Proc 12% (Temperature)	QA 17% (Mixing time)	Maint 22% (Tool wear)	Ops 17% (Cleaning)	Log 17% (Rotation Stock)	Purch 17% (SC Controls Suppliers)
Raw Material Quality					5	5
Dust Control		5	2	3	2	1
Cutting		2	3	3	1	1
Start up routine	2	3	2		1	2
Steady state routine	2		3		1	1
Close down routine	3		3	4		
Total	7	10	13	10	10	10

through the multi-disciplined teams starting with the implementation of low cost or no cost improvements, with further real-time investigation to identify and justify technical improvements.

The X matrix is therefore a means of 'joining up thinking' and grounding the idea of customer value during quality optimisation processes. It tells you where you need to control the process, what regulator shows the performance of the machine and how to ensure that this variable never changes. This is an important step and breaks an old problem that has affected many Western businesses. This issue was caused when, in the old days, continuous improvement groups were established and they cleared away a lot of low hanging issues for operations teams but, over time, the lack of technical skills of the operations teams meant that they could describe problems but did not have the skills to solve them (these were in the heads of technicians and engineers). Soon, continuous improvement became discredited as operators complained and the technical services, which did not hold these issues as high priority, eventually got around to improving the problem. However, by this time, with no real problems that could be solved, the continuous improvement teams either turned into social activities (legitimate skiving) or the programmes collapsed and were discredited. The power of cross-functional teams is therefore enormous and the targeting of improvements must therefore come from the customer and the analyses undertaken by the firm.

Low cost automation

Once stable operation has been achieved, improvement activities can focus on the use of low cost automation. Generally automation which merely replaces human activity will take many years to pay back. Look for opportunities to apply low cost automation to reduce the need for essential non-value-added activities such as inspection, to co-ordinate inter-site material movement and simplify start-up processes. Such programmes should aim to progressively increase in-house capability using a 'learning through doing' approach. Not only does this support the horizontal empowerment approach but it reinforces the important lesson that simplification must precede automation (Table 6.3 sets out five potential levels of automation). Progress through each level adds to the complexity of equipment and the risk of equipment failure. It is important that this is met by an increase in capability. Once stable operation has been achieved, improvement activities can focus on the use of low cost automation.

Table 6.3 Forms of simple automation

		Load machine	Machine cycle	Unload machine	Transfer part
Level of automation	1	Manual			
	2	Manual	Automatic	Manual	
	3	Manual	Automatic		Manual
	4	Manual	Automatic		
	5	Fully automatic			

The loading activity usually contains the highest level of complexity (Figure 6.11). This is where positional accuracy is most critical and where variation in materials can result in jams, and quality problems at upstream processes. Increasingly, this level of complexity is seen as a step too far by many 'world class' companies (including Toyota) and full automation is only ever possible when the last issue for optimisation teams has been solved (the point at which the lights on the business can be turned off and for the machines to work without any form of

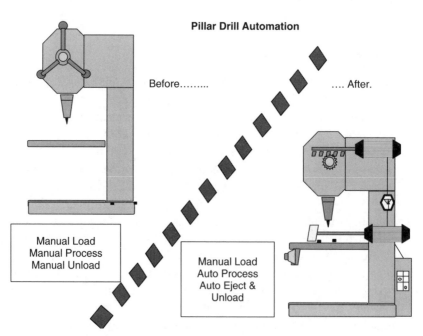

Pillar Drill Automation

Before.........   .... After.

Manual Load
Manual Process
Manual Unload

Manual Load
Auto Process
Auto Eject &
Unload

Figure 6.11 Simple automation applied

interruption). And this brings us to a further capability that is needed to achieve the optimised state and that is optimisation of new assets to be installed at the factory – this pillar is known as Early Equipment Management (EEM) or just Early Management (EM) for short.

6.6 Early Management (EM)

New equipment should be capable of achieving normal conditions from production day 1 (no breakdowns and stable running without unplanned intervention). This is what is meant by the Early Equipment Management goal of Flawless Operations from day 1. Early management is a process for co-ordinating the total design process from product concept onwards. This is important because frequently it is new product development which provides the action trigger for the design/procurement of new equipment and processes. In other words, the product design impacts on the equipment specification that in turn defines the operation. EM takes the lessons learned from taming today's technology in pursuit of greater customer value and incorporates all the improvements to the current technology in the specification of future processes. Quality Maintenance provides the bridge for this knowledge transfer and is a vital source of information when 'designing in flawless' operation and eliminating waste by good design specification.

Early management demands that the inherent design weaknesses of products, services and assets (usually unknown to the asset manufacturer due to the universal rather than bespoke nature of assets) are eliminated and the lessons/failings of the current assets are specified out when seeking to buy the next generation.

Just think how many years are lost to industry each year because central lubrication systems are hidden in the guarded internal mechanisms of metal presses, injection moulding machines and die-casters. If these centralised systems were located on the outside of the asset with no need to spend all that lost production time to access then it would save time and money. This process therefore eliminates the designed in waste from the machine and will include the respecification of the lubrication system (to eliminate the need to stop the asset to lubricate it). But this approach is not limited to lubrication systems redesign alone – EM has enormous power.

It may seem 'overkill' but the ability to specify and 'improve in' the design and standardise the components of new generations of assets will cut a substantial amount of 'proving' and error correction when buying capital equipment. These activities therefore simplify the management of technology and vastly compress the 'ramp up' of new products, services and assets. More importantly still is the collation of knowledge of failures with known assets and, as capital purchases often take a long time, these documents may be used by future generations of engineers and help to eliminate the purchasing of maintenance and operations problems in the future. Too few organisations take this engineering responsibility seriously enough and consequently continue to 'reinvent the wheel'. Again, there are a number of behaviours that limit the potential to create a breakthrough.

What goes wrong?

Managers who are comfortable with a task orientation find it hard to plan effectively. The old carpenter's adage of measure twice cut once is easier said than done when under time pressure. Figure 6.12 sets out the consequences of not doing so. The dotted line on the graph represents the level of input/changes resulting from a traditional project focus. Ideas are sought at the concept stage but resisted at the detailed design stage with the idea of freezing the number of changes. The numbers of

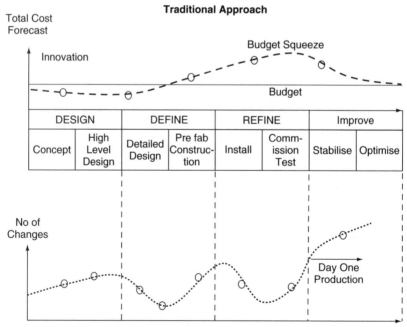

Figure 6.12 The traditional approach to new assets

changes increase at the pre-fabrication construction stage as 'unavoid-able' changes have to be made.

Again, changes are resisted until they are forced through during the commissioning and post-operation activity stages. The top part of the graphic shows the impact on total cost forecasts of late changes. The solid line shows the level of changes identified as part of the EM process. Techniques of visualisation and role definition help to provide access to tacit knowledge, improve communication and take thinking to a deeper level. Experience shows that for similar projects, the area under both change curves is similar.

In other words, the changes generated are roughly in line but the impact on costs is significant. Another point worth noting is that EM techniques actively seek input after the detailed design stage for a number of reasons including:

- It is not possible to specify exactly what a process will look like at the beginning of the project. There needs to be some allowance for a learning curve. Clearly, there needs to be control of costs but costing of every project includes a significant degree of judgement. Trying to pass the risk off to a contractor is a recipe for disaster.
- EM seeks input on innovations to reduce life cycle costs and seeking opportunities to gain added value from the process.

The last point is important. The focus of any design is to deliver the required outcome to 'right size' and 'right spec' assets (that will be around for many years). Once the detailed design is complete, looking for additional benefits from the design is a lucrative source of added value.

The essentials of EM

Figure 6.13 sets out a typical early management review process. This is used to co-ordinate the roles of the three essential partners of the company team mentioned above. They are each accountable on the project for specific targets (Table 6.4). Collectively these six target areas directly impact life cycle costs.

In this framework, production and maintenance are part of the oper-ations function. The roles are designed to provide technical design with the quality of input that they need throughout the project. Below are three project maps setting out the sort of activities involved in a pro-ject to transfer production to a new facility. The commercial focus is on the customer/logistics issues (Figure 6.14) and delivery of the business case, and the same project is shown from the operations perspective

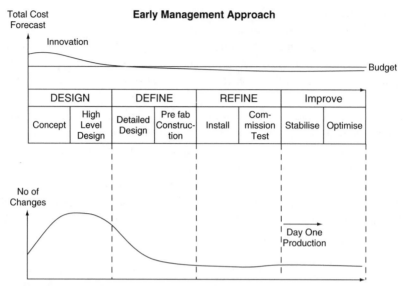

Figure 6.13 The EM approach to new assets

(Figure 6.15) including the development of provisional best practice prior to installation and the application of best practice as part of the commissioning activities. Finally, we have the project from the perspective of the technical team (Figure 6.16). Note the specialist training is left until last. Prior to this the focus is on developing operational competency. Once the plant is running the specialist training can begin to focus on the question of optimisation.

Design to Life Cycle Costs

Life cycle costs provide an important perspective for all stages of the optimisation projects to inform managers of the true cost of owning

Table 6.4 EEM target areas

Function	Target area
Commercial	1 Meeting the voice of the customer. 2 Modelling life cycle costs to maximise business benefit
Operations	3 Operability levels (ease of use). 4 Maintainability (ease of maintenance)
Technical	5 Intrinsic reliability. 6 Intrinsic safety/Environmental

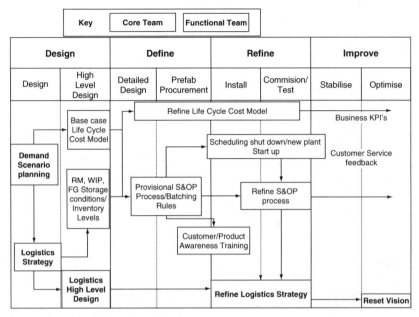

Figure 6.14 EEM – commercial input

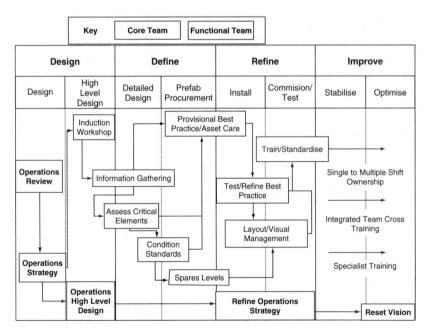

Figure 6.15 EEM – operations input

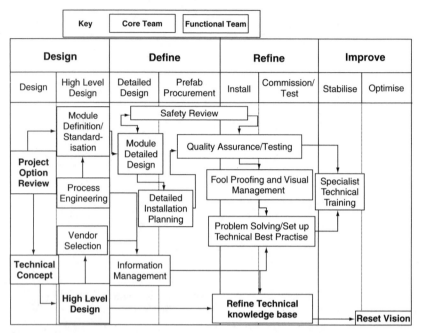

Figure 6.16 EEM – the technical input

the asset base and when certain assets should be replaced and upgraded. Initially the six target areas set out provide a qualitative measure to firstly identify the potential causes of poor effectiveness. At the early stages of the project, you don't know what you don't know. Investing time to review areas of the design where the level of ease of use, maintenance, and intrinsic reliability/safety are low will avoid cost later on. Later on as detail is added and more information becomes available the Life Cycle Cost focus helps to avoid the risk of designing down to a cost rather than up to a value. In some instances the capital cost can be relatively insignificant as shown in Figure 6.17.

Here the gain made by reducing hidden losses were seven times the original capital costs and these benefits would not have been realised without the Early Management Process.

6.7 Capability development

Achieving success with the delivery of optimum conditions needs a combination of technical and people skills/behaviours as set out in Figure 6.18.

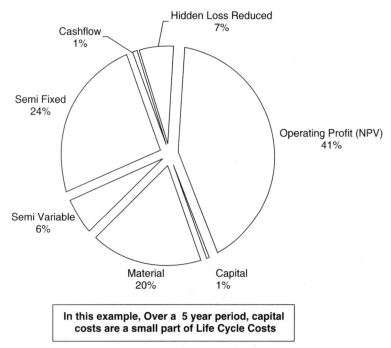

Life Cycle Cost Profile Example

Hidden Loss Reduced
7%

Cashflow
1%

Semi Fixed
24%

Operating Profit (NPV)
41%

Semi Variable
6%

Material
20%

Capital
1%

In this example, Over a 5 year period, capital
costs are a small part of Life Cycle Costs

Figure 6.17 Life Cycle Cost Profile (example)

These are best developed on a progressive basis over one or more projects. Figure 6.19 shows an example of four training modules designed to support this development. When considering the development of Lean TPM skills it is important to revisit the value stream map of the key products manufactured at the factory. These maps clearly

Techniques	Behaviours
• Standard Setting	• Personal Values
• Problem Solving Tools	• Identity
• Information and Analysis	• Principles
• Cause/Effect Mechanisms	• Collaboration Skills
• Value Stream Priorities	• Supportiveness
• Goal Clarity	• Openness
• Team Design/Profiling	• Systems/Process Integration
• Functional Skills	• Learning Process
• Recognition Processes	• Interdependency
• Action/review processes	• Influence

Figure 6.18 Technical peak performances

	TPM SKILLS	Lean Skills	Leadership/Learning Skills
1	Assessing Value Generating Processes, Cause/Effect Mechanisms, Gap Analysis, Strategy Development and Standards Setting	Defining value from the customers perspective, Macro Mapping current and future state	Understanding Leadership, Agreeing future destiny, defining new patterns of behaviour, redefining the rules of the game
2	Goal Clarity, Prioritisation, Accountabilities, Formalise Practices and Problem solving tools	Using process mapping to under-stand where value is added	Setting new priorities and focus, raising standards building empowering rela-tionships, mapping value and action plans
3	Problem Prevention Tools, Cross Functional Processes, Recognition, Action/review process	Improving Process flow	Establishing high per-formance team values and identity to improve new standards, Developing trust and self correcting team processes, conditioning strategies
4	Managing Changes in Working Practices to sustain Zero Breakdowns and Address Optimisation issues	Establishing new ways of working to address the 7 wastes and hidden losses	Managing results not activities, performance review process, the meaning behind performance, using information to energise and maximise productivity

Figure 6.19 Skills for Lean TPM mastery

show where best to invest these skills in project work. The ability to 'see' the impact of improvement activities within the value stream helps determine the commercial benefits of the exercise – a rare phe-nomenon with most improvement activities that don't yield a commer-cial benefit to the firm. Furthermore, the development of the basic Lean TPM skills sets, and then advanced techniques for squeezing the last remaining elements of waste and chronic loss, helps to break through the traditional glass ceilings that have prevented the meaningful involvement of machine operators with process improvement activity. These skills, especially when new products have reached a stable state of order volumes, will be used many times during the service of the individual. Also by creating a critical mass of these skills, the entire class of operators will be capable of sustaining their own improvement efforts. By consequence, this means operators will be able to move to a new level of problem-solving. At this stage these employees can iden-tify (using cause and effect charts) problems with the machines they operate and can select the right tools to analyse these problems in greater detail and then to solve or reduce these barriers to improvement.

Milestone 4: Strive for zero (perfection)

Activities to deliver milestone 4: Strive for zero (perfection) aim to complete the process of changing the competitive landscape to lead the customer agenda for chosen products and services. This builds on the capabilities developed in milestone 3.

Exit criteria includes:

- Delivering industry leading standards of delivery of customer value;
- Established strategy to disrupt current competitive landscape and control the rate of change for other organisations in the same sectors and market segments;
- World Class Standards of Product innovation and Customisation strategies.

An important variable, which must be controlled in order to achieve a high level of material flow, is that of supplier quality assurance. So far we have discussed employees and machinery as inputs to high performance but, for most companies, supplier management routines are poorly developed. But rarely, at any company has it been written that production operators, quality assurance personnel or maintenance engineers should remain in the factory. Indeed, this unwritten constraint in management thinking will slow the achievement of perfection if it remains unquestioned. When all is said and done, these inputs represent materials that happen to be bought in rather than made. As such, suppliers represent 'associated' manufacturing facilities and should be regarded as collaborators in the process of achieving high performance. All too often suppliers are regarded as elements of the production system that should only be treated as 'enemies', to be beaten up for cost reductions and never to have information shared across the two companies so that this information is not used to force higher prices. This view is somewhat outdated (whilst remaining common) by companies engaging in lean practices and the lean approach.

All major Western manufacturers face a growing deficit in the training and quality of new engineers, quality assurance personnel and other employees. This is concerning and potentially creates a vacuum in the supply of new talent to drive the process of 'perfection'. In the future, this means, either manufacturers will have to buy in consulting engineers to help improve processes (even though these individuals don't operate the equipment) or share this talent with customers and suppliers in a form of cross-transfer of people to assist supply chain material flow improvements. This latter activity affects not only the 'door-to-door' Overall Equipment Effectiveness measure but the supply chain effectiveness. Another significance is that, for most businesses, the materials element of product cost is significant (usually 50 per cent or more). When a total costs of supply approach is used, the cost of poor quality, needing to inspect supplied products, the costs of failures at the customers internal value stream and other costs means enlightened business managers will share engineering talent at the very least.

The approach to co-managing the value stream is not new and has been associated with not simply process optimisation but a source of competitive advantage. Here too there are many falsehoods:

1 Suppliers are less sophisticated than their customers;
2 Suppliers should be kept at 'arms length' and prevented from involvement in improvement activities;
3 Suppliers are disinterested in joint improvements with their customers;
4 Multiple sources of supply should be maintained to increase customer power when engaging in price negotiations;
5 Suppliers should be rotated regularly to avoid complacency in the trading relationship;
6 Short contracts should be offered to suppliers to ensure the ability to frequently exert power in price negotiations;
7 Suppliers will use information to increase prices.

The most common approach to integrating suppliers by the major Japanese manufacturers is that of the 'supplier association'. The supplier association is a forum of the key suppliers to the customer and is hosted by the customer organisation in the form of a series of meetings each year used for both sharing and co-ordination of supply chain improvement efforts and also for supplier development. The forum is an important lean mechanism and is therefore used to transfer knowledge between supply chain partners. This knowledge transfer is legitimate because the customer is dependent upon its supply base and in a lean enterprise suppliers represent an important source of innovation and have a direct bearing on the competitive position of the customer organisation.

Putting it simply, there are few better mechanisms than the supplier association in targeting and improving the aggregate performance of productive materials suppliers in the value stream. In parallel, it would be prudent to establish a maintenance association whereby all the chief engineers of important suppliers (rather than commodity suppliers or distributors) gathered together to formally transfer best practices. Forming such a group is important as it leads to better decision-making, a focus on the quality, delivery and cost of supplies, and the integration of suppliers with the main challenges of the customer organisation. Too many Western companies 'shy away' from direct supplier involvement but it is important to understand that, when you purchase products, the optimisation of your production system is dependent upon these suppliers. Under the condition of dependency it is logical to co-operate with suppliers and to investigate the best ways of receiving products and getting them to flow through the production system. Surprising still is the willingness of suppliers to help customers improve and therefore to grow even after years of low supplier involvement and the 'power games' associated with trying to 'get the best price'. In the process of optimisation, supplier involvement is important especially when the goal of the entire supply chain is to optimise material flow and compress the cash-to-cash cycle. It is no wonder that these groups, external quality circles if you like, have been used extensively by 'world class' Japanese manufacturers like Toyota and have been successfully transferred to their Western operations – to the benefit of customer and supplier alike.

6.8 Summary

Building on the success of achieving a stable operation requires a change of outlook at all levels, in particular, the need to move away from action stimulated by operational fire-fighting to that stimulated by a passion for improvement. This process will be frustrating and ultimately unsuccessful unless it is led by a desire to transform the business. As such, change is as much a matter of cultural design as it is technical improvement.

Whilst the technology employed by the factory may be the most sophisticated in the world, it is ultimately the workforce who determines the efficiency of the factory. For employees it is important to take the view that the value of the typical operator must be continuously enhanced so that a lifetime of improvement contribution can be extracted and the quality of working life improved for the individual. Modern manufacturing has less and less room for 'just a pair of hands' and increasingly demands more and more technical/interpersonal skills. The value of employees – the most important source of value to a business and although the mantra that 'employees are our Number

One asset' rolls easily off the tongue, the reality in many factories is that employees remain in a role that has changed little since the 1920s. The elimination of chronic losses and the optimisation of processes (the other assets) are determined by the employees. Even the most up to date machinery will have a reduced efficiency if employees don't know how these assets work and what signs of abnormality can be used to detect failings and take the appropriate countermeasures. Not just that, no business stands a chance of optimising what it does if the skills of the employees are not updated and constantly improved to allow waste to be identified and then reduced. To some this may seem a bit of a waste in terms of giving employees these highly sensitive diagnostic skills but there again these are also the managers who would have believed they had achieved a 'lean business' well before this stage. It can be no surprise that Toyota – the benchmark of all lean businesses continues to promote 'Good Products and Good People' as the basis for its almost 60 years of improvements.

To make this a reality, the top-down leadership challenge is to spring the strategy trap. The bottom-up leadership challenge is to secure delivery of horizontal empowerment. With these dimensions in place, managing the technical optimisation process can then deliver meaningful development of human potential and competitive advantage.

Process optimisation is a never-ending process within the lean enterprise. This is supported by the TPM methodology which combines with TQM to add a reliability of material flow that the lean production system (and TQM) cannot achieve by themselves. High quality and low buffers cannot operate indefinitely without TPM. Lean TPM therefore represents a natural extension of these activities that ensures the minimal level of system buffering (to allow free flowing materials) is supported by a programme of change which focuses on the removal of the inherent failings of the productive asset base. Any future state production system is therefore ransomed by poor or erratic machine performance (especially that of the bottleneck operation). Further, early attempts to optimise the production system will have to contend with existing technology and the TPM optimisation methodology allows both a passage through a human-centred 'glass ceiling' of skills plus a structured means of engineering out the failings of current and future generations of asset.

Process optimisation therefore targets zero losses and the 'zero loss' environment is the end goal of the lean production system within which there are no losses to quality, delivery reliability is assured and costs are at a minimum. These latter features are all, at minimum, 'order qualification' processes that yield customer satisfaction. As such, lean represents a 'why?' businesses are engaging in process optimisation whereas TPM is one of the major 'hows' that the business must engage to create these outputs from the value stream.

The lean approach stresses 'value' as a core business objective. The value of machinery is measured in zero lost time, the elimination of chronic losses and the production of perfect quality products from process capable machinery. However, this is not the entire solution to the achievement of 'world class' performance. The value in closer links with suppliers must be recognised. Buying from cheap sources or businesses that lack the skills to eliminate their chronic production losses will affect the performance of production lines. 'Beating up' these supply sources is hardly a recipe for focusing the minds of suppliers on eliminating waste and enhancing the value they provide. Instead it will result in illogical cost reductions and often the presentation of a price reduction to 'make up for' the lack of improvement activity. A good starting point for suppliers is to show them the waste in their systems by value stream mapping them, using their data to highlight their issues. Furthermore, just because a supplier has an implemented quality management system you should not be fooled into thinking they have the improvement processes or capability to provide you with high levels of customer service, defect-free supplies or a year on year set of improvements.

Bibliography

Hammer, M. (1996) *Beyond Re-Engineering: How The Process-Centered Organization is Changing Our Work and Our Lives*. New York: Harper Collins Business.

7

Sustaining the improvement drive

7.1 Introduction

Implementing Lean TPM is certainly no easy task but it is achievable with hard work, application and making the implementation process something which employees find enjoyable and rewarding. Each stage of the Lean TPM journey includes an element of sustainability – like a ratchet system – whereby gains are secured before moving on to the next stage. The development of a sustainable business improvement model is important if gains are to be recovered over the longer term and the workforce are to feel that additional and self-directed improvements are legitimate without being asked or directed towards areas of improvement. The latter scenario of 'asking' or enforcing improvement solutions has a major drawback in that they are management interventions that tell workplace teams the solutions that must be implemented rather than directing improvement effort in a more general and challenging manner. To this end there are a variety of techniques that allow the self-initiated improvements, sustainability of workplace improvements and factory-wide change processes to take place and sustain (Standard and Davis, 1999).

The issue of sustaining improvement activity lies at the heart of every manufacturing firm, and this most difficult of tasks has two dimensions. The top-down dimension concerns the processes and techniques that can be used at the management level. This is the most fundamental level of sustainability and it is key. All 'world class' manufacturing businesses have a team of committed managers who share a common vision of the future. These organisations know that collaboration with other managers is preferable to conflict (Belcher, 1987). These individuals in the factory control large parts of the manufacturing system and the total supply chain and these individuals occupy positions in the business whereby their decisions can change and optimise the total flow of materials throughout the value stream. To put it another way, several good decisions taken at the management level can generate large changes that release millions of pounds of benefits to the business. Generally, this level of commercial improvement cannot be achieved by operational teams even if they were to offer hundreds of improvement ideas. However, the sustainability picture is neither complete nor effective unless the changes and vision of managers are transferred or deployed to the operational teams that make the value streams of the firm.

7.2   Sustainability at the management level

The sustainability of the Lean TPM approach, at the management level, involves a series of mechanisms which demand integration, learning and consensus building. These are important processes but they take time to develop and must be deliberately scheduled into management meeting time especially where the company managers share a history of 'in-fighting' or 'power gaming' (Peters, 1992).

The first stage of learning concerns two management activities and the first activity is that of engaging as a group in visioning the future and also mapping the current state product value streams. These combined activities allow a focus on 'customer value' to be combined with an understanding of where the company is now thereby identifying the 'value gap'. This 'gap' is therefore the challenge that must be closed if the business is to remain viable. No single business department has all the answers with which to bypass these two processes and therefore in designing a current and future state, virtually all departments will have to contribute to a greater or lesser extent.

From the value stream mapping stage it is possible to identify a wide range of improvement initiatives in each department to improve performance and move the business closer to its 'future state'. These improvements tend to be tactical issues that need to be addressed as they are known or identified weaknesses at the time of the mapping (they remove the barriers that affect today's business). The next stage is therefore to take a view of 'what the customer and business needs' over the next 3 to 5 years. This time period, for determining customer value and also what business stakeholders expect, has many advantages to the modern business.

A 3 to 5 year planning horizon is sufficiently far enough in the future to provide a 'gap' to be closed by the collective action of management but it is not too far ahead to be impossible to predict what is wanted from the business. Focusing this management discussion throws up many interesting interpretations of the future and tends to gravitate to discussions concerning new products, but more importantly, measures of operational performance expected by customers. These commercial discussions, including benchmarking of competitor performance, tend to result in a series of challenges that face the business. Generally these are expressed as a quality improvement target, delivery reliability and lead-time target, and finally a cost-reduction target. As most lean manufacturers know, these targets have a common logic and by improving quality delivery reliability, lead times and costs tend to fall. So the emphasis is most definitely a reconfirmation of 'quality' improvement throughout the business. As these targets and commercial measures pass through every business department (including maintenance engineering) to yield business performance most lean businesses will deploy these challenges to the entire group of middle manager. Middle managers, or departmental managers as they are often called, know what is best practice in their specialisms and therefore they are best placed to find the 'hows' or key projects needed to meet these goals. They also know the changes that must take place within the organisation to deliver these improvements but, under traditional thinking, could not influence these changes.

However, when these middle managers are put in an environment where they are encouraged to find ways of meeting the key business challenges it creates a forum for discussion, learning and also the identification of key improvement programmes. The forum also provides an opportunity to put into perspective the best practices of each department within the context of the overall business performance improvement required. As such the middle management forum allows certain projects, of low commercial worth but regarded as functional best practice, to be de-listed or accepted as a good initiative. Add to these deliberations a finite resource of money or employees and the key initiatives at the departmental and cross-departmental levels will emerge. Having identified these programmes of change it is important to combine them, identify the resources needed and finally to identify a time-line for implementation. There are many techniques that add to this aspect of sustainability and one of the easiest is the 'X' chart form of policy deployment (the name given to taking a set of challenges all the way through to action plans and review processes).

The policy deployment X chart

In Chapter 6 we introduced the X-type chart for product quality attributes and their implications for optimised production management using Lean TPM practices. The same type of chart can also be used to provide a visual profile of business and operations management improvement programme selection. The chart demonstrates the areas in need of improvement and the tasks needed to achieve the business step change in one simple management chart. The format also allows the actual tasks of the improvement activities to be assigned to teams (with a leader and team members) and define the time frame over which the team-based improvement activity will take place.

The X chart is a logical and graphical display that shows the following elements of a focused change process (Figure 7.1). The chart commences with a listing of the key elements of the cost equation and failure sources (Box 1) and then the improvements that the management team have identified as key projects to influence and reduce the total costs of business failure (Box 2). From here the management team determines the targets for improvement (Box 3) and then the savings that achieving these targets would realise (Box 4). Finally, the sum of all the items in Box 4 is presented in Box 5 to declare the overall business savings possible as a result of undertaking the identified initiatives. Returning to Box 2 and the key projects – the far right of the chart shows, for each project, which cross-functional business departments should be involved. It also shows which department has been assigned as the leader of the project and who else will support them. The cross-functional teams are shown in Box 6, and, to the right, a general project time-line showing the start and

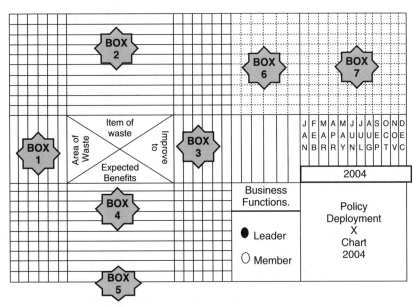

Figure 7.1 The policy deployment X chart

finish of each project (Box 7). The project time-line allows, on a single piece of paper, every project to be seen, its contribution and when it is expected to deliver its value stream improvements.

Figure 7.2 simply shows a couple of projects concerning the management and reduction of 'work in process' costs. Here, two areas of waste have been identified for improvement, batch sizes and set ups, and the target is to halve the batch size and also halve the set up time. Looking across the chart, the set up time reduction programme will be led by operations and supported by the technical department. These two groups of employees will begin the programme of improvement in February and finish it in July 2004 and if they achieve their goal then this will contribute £250 000 of savings for the business. If this round of improvement ideas is continued and all areas of factory waste are identified then it is possible to identify huge savings resulting from only a few projects. As all managers are involved then each will understand the purpose of the project and how they can use the results of the activity for themselves. This is a truly valuable exercise for management, it de-lists those projects that will add no value (or are 'wish list' items for specific managers) and will confirm the major programmes of change needed at the factory level to add more value for the business and its customers.

Having established the key change programmes that have been determined by the cross-functional team to optimise the value stream

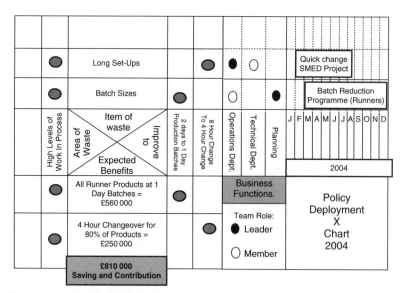

Figure 7.2 Illustration of a business-level X chart (showing projects in operations)

the necessary project management routines need to be introduced. These plans are critical to the business and it is important to understand any linkages between projects that will yield the new production system design (the critical path). At this point, the cross-functional team now has all the analysis and countermeasure projects needed to 'focus' the implementation process and to root out the areas of waste that prevent optimisation of the 'runner' value stream – the one with the most profit potential if redesigned effectively.

Figure 7.2 shows the duration of the project over the coming year or so and identifies who will act as the senior management champion, the operational change leader and the key departments that will form the implementation committee. The latter cross-functional group ensures that all the major stakeholders engage in the change programme and that the 'Lean TPM' initiative is not seen as an 'operations project'. This is an important feature of the optimisation process and this structure offers a very powerful platform upon which to optimise as well as to identify and lower the risks of any given course of action. Furthermore, it also integrates the operations department with customer-facing and supplier-facing business departments in a manner that is focused on exploiting the 'value' of the manufacturing process (for market effectiveness) as much as generating high levels of efficiency.

Iterations of the annual improvement planning process will incrementally identify new levels of organisational waste that constrain the value stream and they will also identify trends and shifts in customer expectations. These are important to ensure a sustainability of process improvements towards the 'zero losses' or 'waste-free' optimised state of production. The 'Lean TPM' approach has a natural complementarity with the needs of other business departments and also prevents the promotion of TPM as merely the concern of the maintenance of production functions and increases the learning across all managers. The latter is an aspect of TPM that is poorly managed in the West even though a successful TPM programme needs a 'total' buy-in from all business managers if it is to yield commercial gains. It is no wonder that traditionally these process improvements have failed to deliver the commercial benefits required when they typically started from an internal rather than customer-focused approach to designing the ideal production system.

The belief of management

Factory management in every business function also has a bearing on sustainability of improvement programmes by their behaviour and attitudes displayed in the factory. This behaviour determines to a large extent the culture of the firm. Any manager who therefore pours scorn on the process of improvement is likely to be heralded (especially by those who seek to slow or hijack the Lean TPM programme) as an example of why employees should ignore the programme and revert to type. Devaluing the skills and competencies needed to fully exploit a Lean TPM programme is a sure way of halting its progress as too is the wanton avoidance of leading change in the factory by those managers. The lean approach therefore needs a personal mandate from the entire management team to move from a traditional organisation to a lean business (Table 7.1).

So therefore a Lean TPM culture must be developed to complement the main improvement and implementation programme. All too often manufacturing businesses ignore this cultural dimension and therefore fail to identify problems with people and miss out on the additional benefits of a culture that accepts change. These issues which will inevitably be faced by managers include a wide range of behaviours which need to be displayed in the factory and written into factory policy to guide the behaviour of managers at every occasion. Also, these policies set a framework for dealing with people – they supplement the formal contract of employment – and can easily be implemented with a little thought. These issues and aspects of a sustainable Lean TPM programme are shown in Table 7.2. Each of these factors requires thought to ensure that intent and reality are matched but no issue is a formidable task and most can be solved with a (or series of) simple solution.

One of the most important aspects of sustainable change within the firm is the regularisation and quality of information passed to the workforce in offices and also at the operations. Even the business costs of replacement parts, lost working time and other financial information should be deployed to these teams. No team can ever know enough about the firm if it is to make the necessary and appropriate contribution to the business. As such these simple issues – the ones that 'close out' management thinking are important to operations level employees. Each of the countermeasures introduced should therefore reinforce the basic values of the firm – that of growth, development and improved value adding. In this respect, many world class businesses feel comfortable with writing down and displaying throughout the factory (and on the business cards of all managers) the formal management policies of the firm. This is good promotion and also serves to 'police' the factory management team. By doing so, it becomes obvious to the workforce when management are backtracking and so must explain why they have been perceived to breach these fundamentals of improvement.

Table 7.1 Old and new management approaches

Traditional business behaviours	Lean TPM behaviours
The business structure resembles highly demarcated functions with rigid hierarchy of responsibility.	Business is oriented to value stream (product family management). A flatter and less layered organisational structure.
Decision-making is centralised to business functions.	Decision-making is decentralised to the value stream team.
Managers plan, instruct and control.	The manager is a coach to the team and develops the team to take on responsibility for processes and their improvement.
Support departments hold all the specialist knowledge in the factory. Knowledge is guarded and not shared.	Specialist knowledge is passed to the team and this transition is overtly managed to allow specialists to join the team or move to higher levels of diagnostic project work (higher value added).
Rigidly written and highly defined job definitions and grades.	Broad job descriptions that increase flexibility of employees. High levels of empowerment.
Employees engage in task activity.	Employees should understand the process, in which they work and should be multi-skilled.
Problems are referred to specialists.	There is no divorce between identifying and solving problems. Employees are appropriately skilled to solve problems to a high quality of solution.
Employee focus is on processing work as fast as possible.	Employees process work as fast as possible given safety and quality requirements of the work and in line with demand for the work.
Focus on cost reduction.	Focus on cost reduction through quality improvement and delivery improvement.
Inspect in quality.	Design in quality and inspection by all.

Table 7.2 Principles and demonstrated application

Key principles	Displayed by:	Why?:
Trust and integrity in the factory	Management behaviour and joint working groups.	Basic condition for workforce collaboration and respect (not necessarily admiration or liking of management).
Mutual respect for co-workers	Secondments to project work. Joint training sessions. Involvement of indirect functions with operations (including sales and marketing).	Basic condition for collaboration.
Openness	Increased use of visual communication within factory. Meetings with appropriate data resources available (not opinion).	Basic condition for collaboration.
Entrepreneurial behaviour of employees	Management–employee briefing sessions and feedback.	Enthuse workforce to feel they can contribute to destiny of factory. Align thoughts to business level not function.
Flexibility of individuals	Broader job descriptions and more salary-style payments.	Less demarcation as a barrier to involvement or rotation within factory.
Customer and market orientation	Communication to workforce concerning business and market conditions.	Promotion of 'value thinking' and benchmarking current activities with a commercial output for the firm.
Teamwork and collaboration	Problem-solving and mapping of processes by cross-functional groups. Common brainstorming sessions. Learning and reflection group time.	Collaboration rather than competition or adversarial relationships.

Less demarcation between employees	Single company uniform, common canteen, common car parking. Wider job descriptions.	Improved flexibility of workers and feeling of legitimate role and involvement with business.
Technical competence	Logical training of employees in skills needed to conduct tasks and problem-solve.	Basic level of management efficiency and effectiveness. Means of deploying decisions down the organisational hierarchy. Decentralises decision-making and stabilises production system by eliminating/identifying errors.
Shopfloor (bottom-up) initiative and involvement. Ownership and autonomy	Performance boards in factory areas. Suggestion process to 'flag up' issues. Clearly marked boundaries and role of each team. Self-managed problem-solving.	
Identification of abnormal conditions in the factory. Quick resolution of problems. Sensitivity to safe working practices	Training in area and asset conditions (normal operating conditions). Auditing of factory standards. Visual management. Regular problem-solving to capture any slippage.	Quicker resolution of problems. Feedback concerning skill shortages in identifying and solving problems.
Formality of standard operating procedures	Written standard operating procedures. Single point lessons.	Basis for improvement so that 're-inventing the wheel' is stopped. Teams build from each other and can share good practice.

Improving performance by measuring it and creating a common language of 'value'

It is often said that if you show a person how they are measured then this will dictate their behaviour (Goldratt, 1984; Kohn, 1999). Thus if middle managers share a common measurement system that includes the quality, delivery and cost of their departments then this is shared by every working area in the factory and appropriate measures of each can be found. It is no surprise that managers use measures to guide sustainable improvements within each department using these measures (and also 'safety' measures and 'morale' measures). In this manner, each area of the factory tends to have a large team performance board (in physical size) that shows line graphs of performance in each of these measures (more often with an annual target of performance for each) (Figure 7.3). In this way, local area teams can use the board to focus their improvement activities and see for themselves the results of their efforts (as the line graphs change). Also, these measures are unlikely to change over the years (instead the target measures will change but quality, delivery and cost will not). This form of sustainability brings together the manager and their teams to talk in a common language and to engage in improvement activity. These boards also form a bond between engaging in Lean TPM activities and witnessing improvements in area performance. At this management-led level, sustainability is personalised to the area of the factory and will provoke 'one to one' interdepartmental working where changes are necessary across shifts or by teams that supply products to the area.

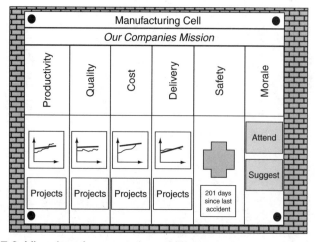

Figure 7.3 Visual performance board illustration

The policy deployment approach and the common measurement system improve communication and add a level of focus to improvement activities that are related to the improvement of the firm. These improvements may also be allocated to individual teams and team leaders in the same manner as that of middle management policy deployment. However, these two elements provide a focus and direction to sustainable change. They are enablers if you like. The implications though include the allocation of time and resources to achieve these improvements – again a feature absent from traditional improvement planning and execution.

7.3 The operations level of improvement

Most people work in the operations level of the firm and it is at this level that an infrastructure and careful attention to sustainable improvements is needed. Our discussion so far has really concerned the supporting practices that allow employees to add value and contribute to the firm. There are many important enablers to sustainable improvement at this level.

At the early stages of a Lean TPM programme, it is important to establish a communication and support mechanism for the factory workers. This is often termed a 'promotion office' and is basically a set of employees who plan, facilitate and guide the improvement programme. These individuals also control the standardisation of factory systems including documentation (especially the quality management system) and training materials. These individuals also have a key role in assessing the quality of the application of improvement activities.

With the role of assessor, the promotion office team has an ability to assist with the necessary training to up-skill the workforce through additional qualifications. The most important qualifications are those of NVQs as these are job-related qualifications that assist operators in gaining a deeper understanding of the processes and technology they control. The ability to combine work with qualifications, especially qualifications that are designed directly to meet the needs of the firm provides an important source of sustainability. Furthermore, it is quite inexpensive as these life skills create a 'pay back' throughout the entire life of the individual's service to the firm. External training also breaks the monotony of the traditional factory environment and signals a clear message to the operations teams that the management is 'serious about change'. These courses also stretch over many months making it difficult to sustain the criticism that 'Lean TPM' is another 'silver bullet' or 'flavour of the month'. Again, there are echoes here of the 'pleasure' of change, reviewed earlier, and rewarded effort.

Another responsibility of the promotion office is to act as a regulator of change (to pace the amount and timing of activities). The ability to control the pace of change is important to 'employee learning' and the achievement of certain standards in the workplace before new techniques will be passed to the teams. At Toyota there is a saying which states 'the teacher will appear when the student is ready' and this is not typical of traditional improvement programmes that teach everything to everyone and expect them to come up with solutions. The logic of the implementation process under Lean TPM is to pace change and reveal the next set of improvements and techniques when there is clear evidence that the last have been mastered. There is another important point to be made here – the promotion office and early stages of Lean TPM are concentrated upon the quality of team solutions and not the quantity. This is important as successful implementation is a learning experience and therefore exemplar innovations must be created (some of these will have low commercial value but will stand in the factory as lessons for other teams). Only when the basics are mastered will an effective promotion office direct efforts towards the quantity of solutions. Such a 'checking' mechanism stops teams from half understanding the tools and techniques and setting off in a direction of their own in an uncontrolled manner.

Add to this approach to capability development the creation of rewards for effort, improvement activities and engagement in key policy deployment programmes and you can stimulate directly a 'learning culture' that is capable of sustaining performance improvements. All too often managers run away from the issue of culture change. It is seen as somehow the last card of resistance or the 'thing' that is most difficult to improve. This is a falsehood: culture is the result of behaviour, and in particular, it is the result of management behaviour. Cynical managers who sneer at improvement programmes will breed a culture of cynicism. Managers, who enthusiastically promote change and improving upon the achievements of today, will not just be happier themselves, but will create a culture that embraces change more readily. In this manner, culture is an output not an input and changing culture means changing the management approach. Think further, treat your trade union as if they are obstacles to progress and guess what they will be, and as a manager you will highlight even the smallest degree of questioning as resistance and guess what, eventually it will turn into resistance. Sustainability of improvements on the shopfloor is therefore a function of sustainability of the approach taken by management. This brings a further point: sustainability of operational improvements is also related to the job tenure of a progressive manager. If there is a high level of management attrition then this will also affect negatively the treatment of 'improvement' by shopfloor teams. Basically, if there is little faith that managers will remain in place to see out the improvement programme then factory teams will take less seriously the mandate for change.

At the operations level of the firm there is a portfolio of practices that support process improvements and these range from giving individuals specific projects to assist their career development to simply acknowledging effort and saying thank you. Further positive reinforcements can be found in terms of offering factory 'certificates' for involvement or achievements following successful change initiatives. These are powerful mechanisms and 'reinforcers' of change, but so too, is the proactive management of getting individual factory employees, or even teams, to attend external training courses and to network with other local businesses. We have even seen complete employees, already good at process improvement, totally reinvigorated after a simple (and free) public visit to a Toyota factory.

Breaking down barriers to change (therefore unblocking improvements) can also be achieved by concentrating on human resource issues such as individual career development plans, succession plans and suggestion schemes. But sustainable improvement is also related to job rotation and getting employees (especially managers) to experience different parts of the production system. This activity extends the boundaries of people's thinking by allowing them to relate their actions to the improvement of other people's performance in the factory. The same is true of operator teams. It is not sufficient to 'bring people in on days' to conduct problem-solving between shifts. These individuals contribute the most when they act horizontally across the business (between teams) and therefore secondment or the deliberate rotation of staff assists in gaining this knowledge and optimising the value stream in a logical manner.

The logic of operations level sustainability

It is astounding that the planning of a major improvement activity, such as Lean TPM, is as we have stated before, the result of 'fag packet' designs and no real attention is put upon the timing and logic of such a change initiative. There is a definite and totally logical approach to the actual programme of events at each factory and this logic is pretty universal. Also within each stage there is a logic that is common to each phase of change.

The logic of Lean TPM implementation is as follows:

1 Create awareness in the factory of a need to change using a collaborative approach to analysing the business as a single system of material flow. This level of lean implementation is best achieved by creating a value stream map by middle managers and key stakeholders in the factory (such as the trade union). The purpose of this stage is not just to create a sense of urgency to gain agreement on what needs to be improved and how best to do it. The 'future state' value stream that results from this activity will inform the business as to how best to get these improvements, what programmes of change are needed and what support is required (including the softer side of employment we reviewed earlier).

2 Implement initiatives to raise worker morale and safety management practices. Here, the CANDO programme of change is invaluable and whilst many texts advocate this as a later stage, we disagree. It is one of the first. It is the first visual sign that management is concerned enough about this latest change programme to engage workers in it (rather than just demanding their compliance with a system they did not build). For some workers, especially ones that have received little formal training by the company or have been a long time away from their school days, this activity is enjoyable and an immense learning opportunity. Furthermore, to implement CANDO successfully it requires an informal level of brainstorming, collaboration and problem-solving. So by doing these individuals gain experience and will be less frightened of classroom-based problem-solving activities.

3 Implement formal quality problem-solving activities with teams. This second stage is superb and complements the CANDO and safety management stage by adding the formal tools necessary to identify, analyse and solve problems. For most employees this stage will be seen as a direct extension of CANDO that is logical to them. For the business, any improvement in quality performance will yield better productivity and therefore lower costs. By engaging in problem-solving the 'collaboration and involvement' aspects of Lean TPM sustainability are refined and exhibited. There is one further feature here that many managers forget – by conducting 'cause and effect' studies the team will identify problems with the environment! These are failings with the CANDO system and can therefore be captured and corrected at this point.

4 Implement new delivery mechanisms. It will take even the most enthusiastic team a while to get to this point and it is here that the

value stream mapping outputs can be implemented at the factory. By engaging in CANDO and problem-solving (plus associated training and other techniques) the management have bought themselves time to get the data and perfect (on paper) the changes needed. Furthermore, the teams have techniques and systems that can be picked up, moved and transplanted in the new production system. At this stage the teams will be involved in establishing the correct procedures and responsibilities to manage pull systems or operate in a flow environment.

5 Reduce costs and improve flexibility of the factory operations. The final stage, at the operations level is therefore to seek out the 'designed in' safety stocks and eliminate them. When companies engage in lean practices they tend, quite rightly, to leave slack in the system that covers for risk. During this stage, after the production system has stabilised, the sources of this slack are questioned. These are costs. Therefore using problem-solving techniques (especially affinity-style diagrams) a causal relationship can be established between improving machine changeovers so as to release safety stocks. This is an important learning activity for the team and extends the tools available to the team and also the level of cross-functional involvement they will seek. Whilst improvements like set up time can be implemented before this stage it is rarely the case that they are regarded as willing additions to the team. More often you hear complaints that 'managers are just trying to make us work harder' and managers will complain that 'operators don't sustain the level of improvements they have made during initial changeover time reductions'. Both these statements exhibit a lack of communication and understanding. This level of change is therefore more potent when it combines a willingness to sustain following self-selection of the problem (not dictation by management).

These general levels are enough to guide the implementation process but within each stage is an important set of criteria that maximise the enjoyment and value of each. Every stage must start with a promotion activity to generate awareness and announce the improvement is coming. This is only common courtesy and also allows some time to get the necessary people released from their 'day job'. Furthermore, even in a factory with traditional resistance to change, it starts the 'rumour mill' going. People begin to talk about it, question it, ask if they can take part and all these behaviours help the change programme.

The logic of implementing the change is also quite straightforward. Teams must find a quality of solution to the problems they face (not quantity). There are many companies that believe in 'blitzing' problems using these events but it rarely sustains. It's the quality of the solution (even to a minor problem) that is more important – quantity comes from repeating the work and sustaining it. So quality first and quantity of improvements second. This approach also firmly states that the company is not interested in meaningless 'short-term' demands for this improvement activity to pay for itself – the repayments will come and they will come for many years if the improvement process is planned carefully.

At the event, the first stage is to generate awareness for the team by showing them good practice and the reality of the factory. Here it is vitally important, especially where certain employees cannot read and write effectively, to train in pictures (show the team!). This avoids problems with getting people talking. It's not that difficult either. Alternatively, take the team to another factory or to sections of your factory that show good practice. At every stage of the improvement programme it is important to build this awareness of the need to change. These sessions will also flesh out what the teams believe the management must deliver to make the programme successful (more often than not its time and resources). The second aspect of sustainability is that of education in more formal terms which includes giving the teams the knowledge and tools to conduct the change and the approach that should be taken. The third stage is to implement the change and monitor its performance (before and after), and finally, the team must meet again to formally review progress. The last stage is important and should not be treated as merely a fifteen minute 'wash up'. Instead, this process should take a couple of hours and include a reflection upon what has been learned, changes that should be made for future events and also to devise the standard operating procedures needed to ensure sustainability when the team project is disbanded. Whilst the project will continue after the team have conducted the stages, these latter activities assist in generating an efficient (minimal time and effort) set of processes that regulate the system after the improvement has been implemented.

These stages apply to all the activity phases from 'safety and morale' to the 'cost reduction and flexibility' stage and to every team. Obviously, repeating the training in the factory with new teams will involve updating and improving the approach and materials used too. The final activity is for the team leader and promotion office personnel to develop a short report (no more than two pages) for submission to the factory management concerning the improvement activity. This report should also include how the team leader expects to continue the programme and whether there were savings (including an agreed diary schedule of future events). These reports form a library of activity logs that can be accessed by other teams facing similar problems. Furthermore, managers who have thought sufficiently about this process will also keep a log of 'who has been trained in what' and the standard they have reached. These are typically displayed in the team area and the promotion office maintains a record. In this manner, trained and experienced individuals can inform others and help them out between formal training sessions (even individuals from other areas of the factory).

This, the logic of sustainability, is therefore to adopt a 'layered approach' to learning so that the new learning points reinforce the last and are seen as a natural extension to what has been done before. Take, for example, the follow-up of a workplace organisation programme (CANDO) with 'Cause and Effect' problem-solving. The Cause and Effect chart has an area of brainstorming related to 'the working environment' and this will pick up any problems that have not been closed out during the CANDO programme. Quick machine changeover programmes also depend upon good workplace organisation and will be seen, by operators, as an extension of the workplace organisation but this time to do with sorting out what is needed at the machine to change it quickly. Layering learning in this manner therefore reinforces what has gone before and is not seen, by the bulk of employees, as just another initiative plucked from the air or from the last conference the managers attended. Such an approach builds an accumulated knowledge in a logical order. You will also notice that these phases actively reinforce the common quality, delivery and cost language and measures of the firm. As such, team performance can be viewed by a cursory analysis of area performance boards and the local measurement systems in place. Obviously, some companies will have effective systems in place already that cover some of these areas but there is no harm in 'refresher' training sessions to reinforce the importance of conducting these tasks properly. Most people work in the operations level of the firm and it is at this level that an infrastructure and careful attention to sustainable improvements is needed. Our discussion so far has really concerned the supporting practices that allow employees to add value and contribute to the firm. There are many important enablers to sustainable improvement at this level.

7.4 Summary

There is no such thing as a 'recipe for sustainability' in terms of a solution that fits all businesses. Even companies making the same products with identical technology will differ in terms of their ability to sustain (anyone who has experienced work in a large divisionalised organisation will appreciate this statement). The recipe is individual and specific to the firm – it is also time-specific and therefore will change especially during conditions where new senior managers move into the firm (this type of strategic management rotation is also a means of preventing sustainability and creating a lack of strategic direction). So companies must decide for themselves what fits and what does not. To some the idea of a common canteen shared by management and workers will be seen as taking things too far – that speaks volumes about the current culture. Sustainability and culture change are not the results of ticking a few boxes on a 'pick and mix' menu of good practices. It takes time but more importantly it takes planning. Rational behaviour and improvement activity must be nurtured, discretions should be identified and highlighted, and persistent employees who actively attempt to derail the programme of change must be dealt with. If they are not then that will become a precedent of defiance and reason why people will not take part in such improvement work. Obviously there is a trade-off in terms of actively seeking questions from employees and seeing these as attempts to subvert improvement. Even some of the brightest people have pretty poor interpersonal skills which to others will come across as offensive or intimidating. That's OK, we are all human but that does not mean these behaviours should not be pointed out to the individual and action taken to address these when they become persistent offences and cause insult. Such events must be handled on a 'case by case' basis but when they compromise the basic principles of the firm, the individual must face a decision to comply with the wishes of the workforce or move on. That is the hard side of sustainability – but it has to be faced. You cannot build a 'world class' business with people who deliberately and persistently defy 'reasonable requests' to change. It is also not fair for the individual.

Having said all this most employees, even hardened cynics, are capable of sustaining change, and most employees enjoy the new skills and variety of working life that Lean TPM brings with it. Further, most would not want to go back to the 'good ole days' – because they simply were not good. At the end of the day the mortgages and holidays of every employee rest upon the performance of the firm so planning for sustainability and autonomous improvements is mandatory. Culture will change as a result – but this happens much more slowly.

Bibliography

Belcher, J. (1987) *Productivity Plus: How Today's Best Companies are Gaining the Competitive Edge*. New York: Gulf Publishing.

Goldratt, E. (1984) *The Goal*. Aldershot: Gower Publishing.

Kohn, A. (1999) *Punished by Rewards*. New York: Houghton Mifflin Publishing.

Peters, T. (1992) *Liberation Management: Necessary Dis-organization for the Nanosecond Nineties*. London: MacMillan Publishing.

Standard, C. and Davis, D. (1999) *Running Today's Factory*. Cincinnati: Hanser Gardner Publications.

8
Conclusions

8.1 Reflections

The foundation of Total Quality Management and the strength of the Toyota-style 'pull system' is only a partial solution to an effective lean manufacturing system. Such a system is inherently weak and no means of competitive advantage in the market place. High levels of quality and low batch sizes are still subject to the disruptions of poor asset management and as such Lean TPM provides an answer to this need. Lean TPM adds a precision and ability to reduce costs whilst reinforcing the drive for improved quality and delivery reliability. It is this power that unleashes the potential to truly transform a business and ensure that the lean ideal of perfect product flow can be achieved and sustained.

Further still, Lean TPM demands a new approach to exploiting the value-adding role of the workforce and operations workers in particular. In many businesses, poorly trained operators with low technical skills control productive assets worth more than a typical Ferrari. It is ludicrous that such a scenario exists and operators are constrained within a 'glass ceiling'. Furthermore, without breaking through this 'ceiling' maintainers cannot achieve their true value-adding potential of stepping up to higher levels of diagnostic and project improvement work. Lean TPM adds together the commercial 'savvy' needed for maintainers with a skills set that is qualitatively different to 'problem-solving'. It moves employees from reactive and event-driven 'restorers' to proactive innovators.

Moving from the 'pain' of traditional 'fix it' type operations to the motivational 'pleasure' of working as part of a winning team, in control and growing in capability is an experience many retired operators and maintainers will have thought impossible. However, as customers become 'more American' and expect the ultimate levels of customer service, the old model will not serve a manufacturing business well. Also whilst manufacturers can do little about the labour cost advantages of other countries, Lean TPM does offer a potent differentiator and a reason to keep Just-In-Time (JIT) local deliveries rather than receive large batches, with long lead times and with quality defect problems from abroad. Providing the highest level of customer service through Lean TPM techniques and the highest levels of process control does make customers more dependent and less willing to take risks with other businesses even though their unit prices may be better.

Few improvement models provide this type of power and benefit to the firm. Lean TPM provides a 'systems approach' involving the entire firm, not a piecemeal 'pick and mix' of independent techniques. Lean TPM is an enabler for new ideas and the application of other successful analysis techniques such as 'Six Sigma'. Such programmes cannot hope to succeed if maintenance and operational performance is poor, erratic and uncontrolled. Simply 'doing' Six Sigma (often motivated by a corporate memo dictating the firm should do so) is an impotent tool in the hands of employees with low technical skills. Doing so will cause frustration especially as no causal relationship will be established between the control of a variable and increased performance when the engineering sub-system of the firm is in chaos. The pity is sophisticated tools, such as Six Sigma, in the hands of 'unstable' manufacturers will inevitably discredit the programme and that is a waste. Further, it takes only one time to discredit a programme and block its relaunch for many years.

On this point, it must be noted that a 'sister' of the Six Sigma approach exists within Lean TPM. Lean TPM is the evolutionary growth of 'asset control' and moving from a chaotic state to stabilisation and then on to the elimination of chronic losses creates a new dimension to maintenance improvement. When companies begin to optimise the control process they enter the frontiers of 'Quality Maintenance' which together with Early Management (EM) provides the bridge between today's zero defect activities to deliver tomorrow's winning product and technologies. This builds on techniques such as reliability-centred maintenance (RCM) and condition-based maintenance (CBM). For example, RCM was used as a core technique within a TPM programme to optimise a refinery operation so that it could achieve 'through the night running without intervention'. RCM with its structured analysis was used to identify those components where only the failure can be detected. This resulted in designing out such components as well as the creation of additional failure curves for corrosion and brittleness to deal with these process industry-specific problems. CBM techniques such as vibration analysis provide an important information in the battle to stabilise and extend component life. It increases understanding of what need to be optimised to deliver zero breakdowns and the ultimate goal of maintenance prevention. Quality maintenance also includes the application of prioritisation techniques such as Quality Function Deployment to better meet the voice of the customer as well as optimisation tools like Taguchi to make defects visible early and proactively control the conditions which reduce their occurrence.

These enhancements are a part of a natural evolution of Lean TPM to deliver a higher level of lean manufacturing performance. It is at this point that statistics becomes more important than rudimentary management statistics used thus far. As such, the Lean TPM system evolves into applied statistics once 'noise' has been cleared from the production system. Call it Six Sigma if you wish, this is certainly one methodology of high value, but typically, maintenance activities will move to predictive activities. Here the key control variables are identified from an engineering perspective, and managed using statistical sampling to predict the imminence of failure.

These advanced tools and methods are therefore best applied to machinery that is a bottleneck or critical to quality performance – and therefore productivity and cost improvements. Often such programmes are introduced before stable operation has been achieved because of the appeal to and to stretch the minds of technical staff – not necessarily operator teams. No amount of statistics can ever replace knowledgeable workers who can flexibly adapt to anything an unpredictable market throws at them.

Without this important organisational capability, technical problem-solving will be less successful and unrecognisable when compared with their application as part of the Lean TPM. They are much more effective when routine activities have been simplified, working relationships are proactive and specialist functions have more time due to the skill and role deployment to the operations teams.

As organisations approach the 'Milestone 4' performance levels, they will focus on sources of failure that lie outside of manufacturing and involve the wastes in the supply chain (poor information exchange, unmet customer needs, synchronised supply chain). The track record of significant improvement will provide the leadership with confidence to grasp new opportunities. It is also highly likely that the leaders involved in this will include personnel in new roles or external recruits brought in to support the move into new areas. It may also include a business expanded through acquisition and incorporate this winning culture into more of the supply chain.

The skills that have helped the company to achieve the highest levels of performance will be the organisations greatest assets in delivering the new challenges and moving onward. With the correct leadership, these skills will never be left to wither or remain unpractised.

Following on from this argument 'what challenges lie ahead for Western manufacturers?' is an interesting question to play with. Traditionally making and designing a product was the area of key concern to most manufacturers. More and more organisations are finding that the line between products and services is becoming less clear: manufacturers who deliver to line side, electricity companies who sell warmth rather than units of electricity, equipment manufacturers who provide finance and insurance deals to name a few. Few areas of competitive advantage last for longer than 10 years. Pharmaceutical products run out of licence, successful innovative products are copied and subject to low cost alternatives, even the retail advantage of prime locations can be reduced. Service differences have an even shorter life. In these conditions developing the organisational ability to lead the customers' agenda is essential if your organisation is to survive the inevitable rationalisation of mature industry sectors and the growth of new ones. This is even more important for publicly quoted companies, where the risk of takeover can only really be countered if the organisation is seen to deliver above average returns.

In this book we have written with the basic assumption that manufacturing businesses are making the right products combined with the right service package. Lean TPM, with its focus on customer value, addresses the issue of making products customers really want to buy together with the services to increase its value to them or deliver win/win reductions in supply chain costs. As such, the next generation of Lean TPM improvements is likely to see a greater bond of the engineering talent in an organisation with the corporate power brokers in the marketing department to deliver the right product/service package. There can be nothing worse that making the wrong product in the most efficient way. So it is highly likely that 'marketers' (whose mantra is 'to be close to the customer') will seek to promote the environmentally friendliness of product designs and also to seek new solutions to customer needs.

This process, which traditionally was hands-off and designs were thrown into the operations to be made, will naturally lead to greater 'design engineering' and 'time-to-market' initiatives. The greater quality of information from the customers will provide opportunities to capture innovation at all levels. As innovation is the art and science of making linkages between ideas which already exist but are not currently recognised, the use of innovation methodologies will become more widely used. These do not replace raw creativity but they make the process more productive and provide a practical process for involving the whole organisation in thinking outside of the box.

As we have argued many times in this book, unless the Lean TPM programme has resulted in a stable and controlled production process, then involvement at this level of business decision-making is pretty meaningless. Also, the heads of engineering and operations departments are unlikely to get this involvement just by pointing out the logic of this integration or that 'world class' companies do it. No operations manager can achieve meaningful integration with the design process unless current products are made in a stable and controlled manner. Also of importance is that the integration with product design offers the ability to influence the engineering of the product, the need to analyse competitor products (strip downs) and a greater involvement in the costing of future products. These skills are not typically used by engineers, and proficiency in these activities places engineering at the centre of all strategic business discussions. Generations of engineers have sought this elevated management position and despite this lesson having been demonstrated by Japanese and exemplar manufacturing performers throughout the world, the West has been slow to understand this lesson and truly release engineering as a competitive weapon.

Another interesting direction for Lean TPM companies is the transfer of knowledge and practices to the strategically important suppliers to the business. Nowhere is it written that engineers and operators cannot be actively involved in the development of production and maintenance suppliers. This is purely a mental barrier that does not exist in reality! Just as boundaries between business departments are purely lines on an organisation chart. This opens the gates to a wide range of activities that will call for a new skill set, one that is capable to taking Lean TPM tools and techniques into businesses with very different technologies. This new opportunity may seem unrealistic but as Western Europe, and Britain in particular, faces a widening deficit of trained engineers from quality universities and colleges, this ability to transfer skills will be at a premium. Predictions already show that youngsters would rather enter the media not engineering, students aren't taking the sciences necessary to get them to college and 'non-completers' of engineering degrees continue to rise. Also, as engineering skills rely upon a good diagnostic capability, these individuals are being targeted by service industries and thus qualified engineers are being attracted to these businesses and away from adding value to the manufacture of physical products. So any business that limits its engineers and capable operations staff to within the business may be missing a trick in terms of improving the performance of its supply chain. This many effectively serve to ransom the ability of the company to exploit its manufacturing potential as supplier problems cause 'stop–starts' to the internal value stream.

For companies that engage the Lean TPM process, the future will be full of learning opportunities and a series of watershed points. At these watershed points the limiting factors within the business, its policies, approaches to human resources and technological innovation processes will be incrementally broken and reconstituted to meet the strategic demands of the business. These experiences are valuable and represent transitions for the firm as it reshapes its sources of competitive advantage. Further still, experience shows that a world class manufacturer is supported by world class suppliers. This virtuous circle does not come about by chance but by design. Any business that thinks it can simply 'magic up' a good supply base is misguided. It will not happen.

Any company that thinks it can just switch sources to better ones is also misguided as a supplier's willingness to help you achieve a world class level of performance is based upon the supplier's belief that you will help them achieve it too. To put it another way, you need a good relationship with your suppliers. This is not that surprising and the key to a world class level of supply chain OEE performance is collaboration. Under these conditions, suppliers are regarded as extensions of the basic factory system. They are members of the production system conducting tasks that the company does not or cannot do by itself. Collaboration is therefore vital and this means breaking down even more barriers – ones external to the business. These barriers don't exist in reality but they are invented barriers in the minds of managers. So working with suppliers and sharing resources is quite a natural activity if you adopt the mindset that it is better to collaborate with suppliers than compete with them. Furthermore, it is when collaboration occurs that suppliers become comfortable in declaring their manufacturing costs in a transparent manner to customers and it is here that 'exchange-engineering' can improve the profitability of the supplier whilst lowering the purchase cost of materials to the customer firm (the employer of the engineer).

So whilst the economic climate is challenging to British manufacturing industry there are major opportunities to improve using the Lean TPM approach. Silver bullet it is not. The biggest single change programme and the most important business investment made by your company in the past 20 years it certainly is.

We hope that we have convinced you to follow the powerful Lean TPM blueprint for change – a whole suite of enjoyable experiences now await you.

Index _____

TPM5, 24
Transforming the business model:
   five stages, 88–89
   logic, 85

Value stream, visualisation, 115–19
Voice of the Customer (VOC), 110,
   113–15

Welch, J., 52
Willmott, P., 47

Womack, J., 19, 24, 26, 27, 29, 31, 47,
   55, 63, 85, 97, 110, 125
World class manufacturing
   model, 17, 159

X chart planning, 123, 147–9, 166–9

Yamashina, H., 128

Zero losses, 4, 141